Geology of the country around Garstang

The continuing search in Lancashire for natural resources such as oil and gas, and aggregate materials for the construction industry (sand, gravel and limestone) highlights the need for up-to-date geological information. Geological data are also essential for planners and civil engineers in a district traversed by major arterial road and rail routes as well as vital oil, gas and water pipelines. This memoir is intended to supply this basic need and to indicate where more detailed information exists in the Survey's extensive archives.

The district described in this memoir extends from the uplands of the Bowland Fells and Longridge Fell, in the east, to the coastal plain of the Fylde, in the west. The rocks at outcrop in the upland area range in age from early Carboniferous limestones and mudstones, laid down in tropical seas, to late Carboniferous sandstones and mudstones, deposited by rivers that flowed into the sea from the north and east. Large faults separate this area from the Fylde lowland, where a Permian and Triassic redbed sequence, formed in near-desert conditions, overlies the Carboniferous rocks. Uplands and lowlands alike were glaciated during the Pleistocene ice ages, which ended about 10 000 years ago. The deposits of the last ice age were deposited thickly in the Fylde and more patchily elsewhere, obscuring the nature of the bedrock in many places.

The detailed studies summarised in the memoir throw light on the development of the early Carboniferous Craven Basin, which has shown some hydrocarbon and base metal mineral potential, and on the later deformation of the Carboniferous rocks during the evolution of the Irish Sea Basin. The Permo-Triassic rocks of the latter contain salt deposits and hydrocarbon accumulations in adjacent districts, and include an important aquifer, the Sherwood Sandstone, throughout the Fylde region.

Cover photograph

View looking south down the valley known as the Trough of Bowland, situated in the heart of the Bowland Fells in the north-eastern part of the district. The Bowland Shales underlie the rushy slopes in the foreground and encircle the Dinantian limestones forming the core of the Sykes Anticline exposed in the disused quarries visible on the righthand side of the valley in the middle distance. Totridge Fell, formed of Pendle Grit, is in the background. (A14809)

Frontispiece   The Vale of Chipping and Bowland Fells viewed from near Cardwell House on the Longridge Fell escarpment.

The vale is floored mainly by till-covered argillaceous rocks of the Bowland Shale and Worston Shale groups overlooked by the Pendle Grit escarpments of Longridge Fell and the Bowland Fells, notably Parlick and Fair Snape Fell to the left and Totridge to the right (A14798).

BRITISH GEOLOGICAL SURVEY

N AITKENHEAD
D McC BRIDGE
N J RILEY and
S F KIMBELL

Geology of the country around Garstang

Memoir for 1:50 000 geological sheet 67
(England and Wales)

CONTRIBUTORS

Seismic interpretation
D J Evans

Petrography
B Humphreys

LONDON: HMSO   1992

First published 1992

ISBN 0 11 884485 7

Bibliographical reference

AITKENHEAD, N, BRIDGE, D McC, RILEY, N J, and
KIMBELL, S F. 1992. Geology of the country around
Garstang. *Memoir of the British Geological Survey*, Sheet 67
(England and Wales).

Authors

N Aitkenhead, BSc, PhD
D McC Bridge, BSc
N J Riley, BSc, PhD
S F Kimbell, BSc
British Geological Survey
Keyworth

Contributors

D J Evans, BSc, PhD
B Humphreys, BSc
British Geological Survey
Keyworth

Printed in the UK for HMSO

Dd 295252   C10   06/92

Other publications of the Survey dealing with this and adjoining districts

BOOKS

Memoirs
Geology of the country around Blackpool, Sheet 66, 1990
Geology of the country around Preston, Sheet 75, 1963
Geology of the country around Settle, Sheet 60, 1988
Geology of the country around Clitheroe and Nelson, Sheet 68, 1961
The geology of the Rossendale Anticline, Sheet 76, 1927

British Regional Geology
The Pennines and adjacent areas, 3rd Edition, 1954

Economic Memoirs
Geology and hematite deposits of South Cumbria, Sheet 58, 1977
Geology of the Northern Pennine Orefield: Volume 2, Stainmore to Craven, 1985

Mineral Assessment Report
No. 116   The limestone resources of the Craven Lowlands, 1982

MAPS

1:625 000
Great Britain South, Solid geology, 1979
Great Britain South, Quaternary geology, 1977
Aeromagnetic map (south sheet), 1965

1:250 000
Liverpool Bay, Solid geology, 1978
Liverpool Bay, Seabed sediments and Quaternary geology, 1984
Liverpool Bay, Bouguer gravity anomaly, 1977
Liverpool Bay, Aeromagnetic anomaly, 1978
Lake District, Solid geology, 1980
Lake District, Seabed sediments and Quaternary geology, 1983
Lake District, Bouguer gravity anomaly, 1977
Lake District, Aeromagnetic anomaly, 1977

1:50 000 or 1:63 360
Barrow in Furness (Sheet 58) Solid with Drift, 1976
Blackpool (Sheet 66) Solid with Drift, 1975
Southport (Sheet 74) Solid with Drift, 1989
Garstang (Sheet 67) Solid, 1990
Garstang (Sheet 67) Drift, 1991
Preston (Sheet 75) Solid, 1982
Settle (Sheet 60) Solid, 1988
Settle (Sheet 60) Drift, 1991
Clitheroe (Sheet 68) Drift, 1960
Rochdale (Sheet 76) Drift, 1974

1:63 360
Preston (Sheet 75) Drift, 1940
Clitheroe (Sheet 68) Solid, 1960
Rochdale (Sheet 76) Solid, 1927

1:25 000
SD74 and part of SD84 Clitheroe and Gisburn, Solid with Drift, 1970

CONTENTS

FIGURES

PLATES

TABLES

PREFACE

The continuing search in Lancashire for natural resources such as oil and gas and aggregate materials for the construction industry — sand, gravel and limestone — highlights the need for geological information. Such information is also essential for planners and civil engineers in a district traversed by major arterial road and rail routes as well as vital oil, gas and water pipelines. This memoir, the first account of the geology of the district around Garstang ever to the published, is intended to supply this basic need, and to indicate where more detailed information exists in the Survey's extensive archives. It is best read in conjunction with the colour-printed 1:50 000 scale published geological map.

The district described extends from the scenically attractive uplands of the Bowland Fells and Longridge Fell, in the east, to the coastal plain of the Fylde, in the west. The rocks at outcrop in the upland area range in age from early Carboniferous limestones and mudstones, laid down in tropical seas, to late Carboniferous sandstones and mudstones deposited by large rivers that flowed into the sea from the north and east. In the Fylde lowland, a Permian and Triassic redbed sequence, formed in near-desert conditions, overlies the Carboniferous rocks. Uplands and lowlands alike were glaciated during the Pleistocene ice ages, which ended about 10 000 years ago. The deposits of the last ice age were deposited thickly in the Fylde and occur more patchily elsewhere, obscuring the nature of the bedrock in many places.

The comprehensive studies summarised in the memoir incorporate the results of detailed geological mapping by the British Geological Survey, as well as subsurface information from various site investigations, and from exploration by the metalliferous minerals, petroleum and water industries.

The synthesis of this information presented here throws light on the development of the early Carboniferous Craven Basin, which has shown some hydrocarbon and base metal mineral potential, and on the later deformation of the Carboniferous rocks during the evolution of the Irish Sea Basin of Permo-Triassic rocks; the latter contains salt deposits and hydrocarbon accumulations in adjacent districts, and includes an important aquifer throughout the Fylde region.

I am confident that this memoir will play its part for many years to come, not only in the probable discovery and exploitation of new mineral resources, but also in the necessary planning and conservation of this attractive district in the interests of the community.

Peter J Cook, DSc
Director

British Geological Survey
Keyworth
Nottingham NG12 5GG

16 March 1992

ACKNOWLEDGEMENTS

NOTES

In this memoir, the chapters on Dinantian rocks, Quaternary deposits and geophysical investigations were written by Dr N J Riley, Mr Bridge and Mrs S F Kimbell respectively. The remaining chapters were largely written by Dr Aitkenhead, who was also responsible for compiling the memoir. Dr D J Evans contributed to the chapter on structure, and the results of petrographical work on some of the Carboniferous rocks were provided by Mr B Humphreys. Data from the BGS Technical Reports written by Mr R G Crofts and Drs T P Fletcher, A S Howard and R A Hughes (listed in Appendix 3) have been freely used in this account. Dr N J Riley identified all the Carboniferous fossils except the miospores, which were determined by Dr B Owens. The photographs were taken by Messrs T P Cullen and C J Jeffrey; a complete list of available photographs is given in Appendix 4. The memoir was edited by Mr J I Chisholm.

We gratefully acknowledge information and assistance generously provided by North West Water and their predecessors for borehole and hydrogeological data, Lancashire County Council for motorway site investigation data and B P Minerals International Ltd. for mineral exploration data. We also acknowledge help given by Gas Council (Exploration) Ltd, Pendle Petroleum Ltd, Tarmac Roadstone Ltd and Dr A Sims. In addition, we acknowledge the access and help given by numerous farmers and landowners throughout the district during the course of the geological survey.

The resurvey was supported in part by the Department of Trade and Industry and the North West Water Authority.

Throughout the memoir the word 'district' refers to the area covered by the 1:50 000 geological sheet 67 (Garstang).

National Grid references are given in square brackets; all lie within 100 km square SD.

Numbers preceded by the letter A refer to the BGS collection of photographs; those preceded by the letter E refer to the BGS sliced rock collection.

The authorship of fossil names is given in the fossil index.

Enquiries concerning geological data for the district should be addressed to the Manager, National Geosciences Data Centre, Keyworth.

HISTORY OF SURVEY OF THE GARSTANG SHEET

The district covered by the Garstang (67) sheet of the 1:50 000 geological map of England and Wales was originally surveyed on the six-inch County Series sheets Lancs 40,44,45 and 53 by C E de Rance, R H Tiddeman and J Shelswell, and published on one-inch Old Series sheet 91SE in 1883. In 1971, an area east of Garstang was resurveyed on the six-inch scale by Messrs J I Chisholm, W B Evans and Dr N Aitkenhead. The remainder of the district was resurveyed on either the six-inch or 1:10 000 scales in 1983–87 by Messrs D McC Bridge, R G Crofts and D Price, and Drs N Aitkenhead, T P Fletcher, R J O Hamblin, A S Howard and R A Hughes.

Geological 1:10 000 scale National Grid maps included wholly or in part in 1:50 000 sheet 67 (Garstang) are listed below, together with the initials of the geological surveyors and dates of the survey; in the case of marginal sheets, all surveyors are listed. The surveyors were: N Aitkenhead, R S Arthurton, A Brandon, D McC Bridge, J I Chisholm, R G Crofts, J R Earp, W B Evans, T P Fletcher, R J O Hamblin, A S Howard, R A Hughes, E W Johnson, R C B Jones, E G Poole, D Price, L H Tonks, T H Whitehead, A J Whiteman and A A Wilson.

Copies of the fair-drawn maps have been deposited in the BGS libraries at Keyworth and Edinburgh for public reference and may also be inspected in the London Information Office, in the Geological Museum, South Kensington, London. Copies may be purchased directly from BGS as black and white dyeline sheets.

SD 33 NE	WBE	1968
SD 33 SE	WBE	1968
SD 34 NE	AAW	1968
SD 34 SE	AAW, RCBJ, THW	1934, 1968
SD 43 NW	DP	1985
SD 43 NE	DP	1985
SD 43 SW	RCBJ, THW, DP	1933–34, 1985
SD 43 SE	THW, DP	1933–34, 1985
SD 44 NW	RGC	1985
SD 44 NE	NA, RGC	1971, 1986
SD 44 SW	RGC	1985
SD 44 SE	NA, RGC	1971, 1986
SD 45 SW	RGC	1985
SD 45 SE	RGC	1986
SD 53 NW	DMcCB	1986
SD 53 NE	DMcCB	1985
SD 53 SW	LHT, THW, DMcCB	1933–35, 1986
SD 53 SE	LHT, DMcCB	1933–35, 1986
SD 54 NW	JIC, WBE, NA	1971, 1987
SD 54 NE	ASH	1986
SD 54 SW	NA	1971, 1983
SD 54SE	NA	1984–85
SD 55 SW	JIC, AAW, ASH	1971, 1980, 1984, 1986
SD 55 SE	RAH, EWJ, AB	1984–86
SD 63 NW	RJOH, DMcCB	1984–85
SD 63 NE	JRE, DMcCB	1949–50, 1985–86
SD 63 SW	DMcCB	1986
SD 63 SE	DMcCB	1986
SD 64 NW	ASH	1985–86
SD 64 NE	JRE, EGP, AJW, TPF	1950–53, 1985
SD 64 SW	NA	1985–86
SD 64 SE	JRE, EGP, AJW, TPF	1950–53, 1984
SD 65 SW	RAH	1985–86
SD 65 SE	EGP, RSA, TPF	1953, 1979, 1985

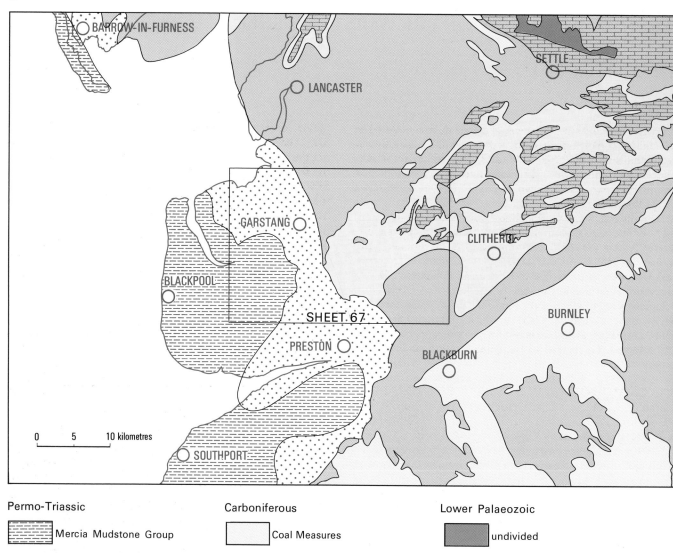

Figure 1 Sketch map showing the general geological setting of the district.

ONE

Introduction

This memoir describes the geology of the district covered by the 1:50 000 Garstang Geological Sheet (67) published in solid and drift editions in 1990 and 1991 respectively. The district lies within the county of Lancashire and is almost wholly rural in character, with much of the land being used for pastoral farming.

The district is clearly divisible into two contrasting parts, each with topography directly related to the geology (Figures 1 and 2). In the west, a low-lying area of subdued relief known as the Fylde is underlain by relatively soft, drift-covered, Permo-Triassic sandstones and mudstones. To the east, more diversified upland scenery formed from Carboniferous rocks includes the high moorlands of the Bowland Fells* and Longridge, where tough sandstones are at outcrop, and intervening areas of low ground, such as the Vale of Chipping (Cover picture) and the Hodder valley, where

softer mudstone-dominated parts of the sequence are present. The scenery in the last-named area is enhanced by large limestone knolls and abrupt transitions from moorland to wooded valleys, which reflect the local complexity of the geological structure (Plate 3). The Fylde was, until recently, subject to extensive flooding. Hence all the major south–north through routes including the M6 motorway follow the eastern margin of the plain.

While the solid geology determines the broad features of the topography, the drainage pattern and the individual character of the river valleys are strongly influenced by the last glaciation and its deposits, especially the thick spread of

*This commonly used term is preferred in the present account in place of the ancient but potentially misleading 'Forest of Bowland' which appears on Ordnance Survey maps.

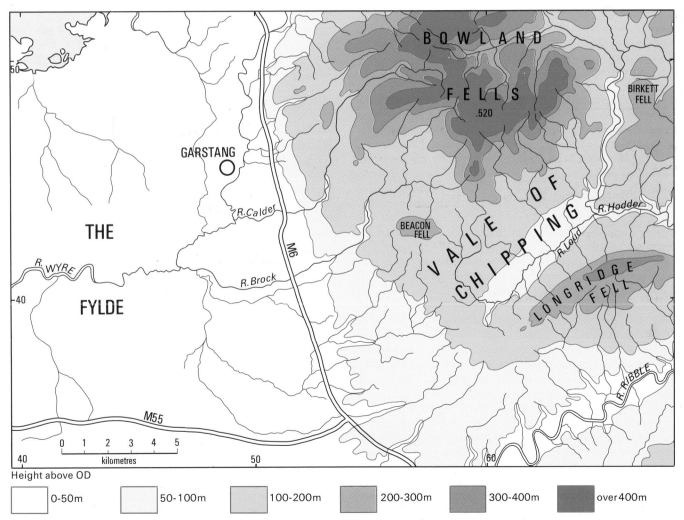

Height above OD

0-50m	50-100m	100-200m	200-300m	300-400m	over 400m

Figure 2 Principal physical features and drainage.

till. On the Fylde, however, extensive spreads of river and marine silt overlie the till sheet and pass almost imperceptibly, north-west of Winmarleigh, into the shifting tidal sands of Morecambe Bay. The layer of peat that once covered the low-lying flats has now been largely removed. In places, buried peats attest to the variations in sea level that have taken place in postglacial times.

OUTLINE OF GEOLOGICAL HISTORY

The oldest rocks proved in the Garstang district are limestones of early Carboniferous age. These are believed to rest unconformably on strongly deformed Lower Palaeozoic rocks, as in the nearby Settle area some 22 km away to the north-east (Arthurton et al., 1988). Old Red Sandstone rocks may also be present above this unconformity.

Following the deformation of the Lower Palaeozoic rocks, and towards the end of the Devonian period, some 374 to 365 million years ago (Forster and Warrington, 1985; Leeder, 1988), much of what is now Britain probably lay in the southern part of the Old Red Sandstone or Laurasian continent, which was at that time situated in the equatorial belt south of the equator (Johnson, 1981, p.221; Johnson and Tarling, 1985). Far to the south lay the northern margin of the huge continent of Gondwanaland, which was to move relentlessly northward until the two continents were fused in the climax of the Hercynian (Variscan) orogeny at the end of Carboniferous times. Leeder (1982, 1988) has summarised the evidence and suggests that, long before the final compressive movements, the Laurasian continental foreland or back-arc terrane was stretched by the flow of material in the hot mantle layer underlying the crust. The stretching produced rupture in the form of a series of tilted fault blocks, one of which is thought to have formed the foundation for the Craven (or Bowland) Basin, where much of the Carboniferous sequence of the present district was deposited (Gawthorpe, 1987). At an early stage in its formation, this tilt block was itself disrupted into smaller blocks (Lee, 1988b, fig. 8.3). Since this probably occurred while the region was sinking beneath the sea, great thicknesses of sediment accumulated locally in downthrown areas. There may also have been some lateral strike-slip movements along pre-existing lines of weakness in the basement. Such movements cause local folding, which can also affect sedimentation (Arthurton, 1984).

During the middle and late Dinantian (Chadian to Brigantian stages) a muddy open marine environment prevailed in the district. However, deposition of sediment was greatly affected by the interplay of intermittent basement faulting, varying rates of subsidence in response to regional crustal sag, and eustatic change of sea level.

Thus, at times, parts of the basin were swept by turbidity currents transporting mainly carbonate sediment, either from up-faulted shallower areas (highs) developed within the basin or from marginal shelf areas. During more stable times sedimentation tended to catch up with subsidence and more uniform shallow-water environments prevailed. These conditions favoured the growth of marine organisms whose comminuted calcareous remains made important contributions to the accumulating sediment. During one interval in the

late Chadian stage, localised buildups of organically generated lime mud (knoll-reefs) developed, greatly enhancing the sea-floor relief. There were also rare occasions when erosive events occurred, due either to local uplift or to fall in sea level. The products of these events accumulated locally as fans of coarse carbonate debris.

Early Namurian times brought a change to a wetter equatorial climate as a result of shifts in atmospheric circulation due to the relative movements of continents and oceans (Rowley et al., 1985). Additionally, perhaps, there were great physiographical changes in the hinterlands from which the sediments were derived. River systems, whose deltas had occasionally brought minor influxes of sand to the basin in Dinantian times, became rapidly very powerful and extensive, bringing great quantities of sand into this part of the Craven Basin. This epoch, like the earlier one, also saw the repeated waxing and waning of ice sheets in the southern hemisphere. The glacial oscillations were probably reflected in the cyclical (eustatic) changes in sea level, that repeatedly swamped the deltas, cutting off the sand supply and allowing fully marine conditions to be re-established in the basin. After repeated temporary setbacks of this nature, the prograding delta top eventually established itself in the Garstang district, probably in mid-Namurian (?Alportian) times.

Rocks deposited during the next time interval, spanning the later part of the Namurian and most of the Westphalian, are not present in the district. However, it is estimated that about 3.5 km of these rocks accumulated, but were eroded near the end of Carboniferous times, 300 million years ago, when the main Hercynian compressional movements, the final effect of the collision between Gondwanaland and Laurasia, resulted in strong folding and uplift.

Trapped between the grains of sediment in the deeply buried Dinantian and early Namurian rocks were organic material and sea water containing dissolved minerals. These sediments, underlying the postulated pile of late Namurian and Westphalian rocks, were subjected to increased temperatures and pressures, which induced chemical reactions in the intergranular constituents to produce fluids containing such metals as calcium, barium, lead and zinc, together with hydrocarbons. These fluids migrated to suitable porous rocks, where minerals and hydrocarbons were deposited. Although no economic deposits have yet been discovered in the district, intensive exploration by oil companies continues.

By early Permian times the Hercynian movements had died down and the landscape formed out of the deeply eroded Carboniferous rocks probably had a relief not unlike that of the present day. The climate was hot and dry, the exposed land surface became deeply oxidised and reddened, and the valleys were filled, mostly with sand. A new arm of the sea then extended over the western part of the district and red mud with subordinate dolomitic carbonates and minor evaporites were deposited in it. At the same time a new phase of north–south-oriented rifting movements began, and these were to dominate and control Triassic palaeogeography and sedimentation in the western half of Britain during the Triassic Period. Rifting produced fault-bounded basins into which an extensive river system brought great quantities of sand. This influx of sediment came from the rejuvenated Armorican mountains far to the south, in what is now the English Channel and Brittany. The river system gradually lost its transporting power and by middle Triassic times, when the Mercia Mudstones were deposited, the subdued landscape was probably one of broad floodplains and ephemeral lakes, at times briefly connected to the sea and at other times drying out, causing salt to be deposited.

There is no record of events in the Garstang area from middle Triassic times, about 225 million years ago, until late in the last glacial period, about 20 000 years ago. Almost certainly there were phases of deposition, during major global marine transgressions in Jurassic and late Cretaceous times, and phases of erosion, consequent on general uplift and doming of the Pennine axis during the latest Cretaceous to early Miocene interval. There is also evidence from cave deposits, in the nearby Settle district, that large mammals such as elephant, rhinoceras, hippopotamus and hyaena liv-ed in the region in the warm Ipswichian Interglacial that immediately preceded the last (Devensian) glaciation (Gascoyne et al., 1981).

This glaciation, at its maximum, completely buried the landscape in ice, which moved across the district from source areas to the north and north-west, and from the Pennine iceshed to the north-east. It left its erosional imprint on the topography and also covered much of the district with a blanket of drift whose variable lithology includes clayey melt-out tills and the generally more sandy sediments of glacial meltwaters.

During the postglacial (Flandrian) stage, a new drainage pattern was established, sea level intermittently rose and the present-day processes of erosion and deposition became established, increasingly modified by human intervention.

NA

TWO

Dinantian rocks

Dinantian rocks underlie the entire district and crop out in a series of north-east-trending folds along the Vale of Chipping and into the Hodder Valley (Figures 2 and 5). They are the oldest strata exposed in the district. In the north-west, beneath the Bowland Fells, and in the south-east, beneath Longridge Fell and in the Ribchester area, the Dinantian rocks are concealed by the Silesian outcrop. In the west, the Dinantian sequence is faulted against and overstepped by Permo-Triassic rocks. Over much of the central and western part of the Vale of Chipping the outcrop is concealed beneath glacial drift.

During the recent resurvey, the Dinantian rocks were mapped and described by Fletcher (1987, 1990, 1991), Aitkenhead (1990a, 1990b), Hughes (1986), Bridge (1988b, 1988c, 1989) and Howard (in preparation, a). Some of the findings of these geologists have been incorporated into the account which follows.

The limestones in the sequence are generally impure, although the local impersistent "reef limestones" can comprise thick sequences of relatively pure limestone. A significant proportion of the succession comprises rocks of terrigenous provenance, including mudstones or siltstones and, less commonly, sandstones.

The exposed thickness of the Dinantian succession is around 1.5 km. Geophysical evidence (Chapters six and seven) suggests that, locally, a maximum of 4.5 km is concealed at depth. It is not known whether the Dinantian hereabouts is directly underlain by Devonian strata or older rocks (p.79).

PALAEOGEOGRAPHY AND DEPOSITIONAL HISTORY

The north of Britain, during the late Devonian and Dinantian, was subjected to crustal stretching (Leeder, 1982, Gawthorpe et al., 1989). This resulted in a series of extensional, fault-bounded, rapidly subsiding basins, separated by slowly subsiding horst and tilt-block areas (Miller and Grayson, 1982). This tectonic regime had a profound effect on deposition, resulting in a relatively complete, usually thick argillaceous sequence in the basins and an incomplete, usually thin limestone sequence elsewhere. The present district (Figure 3) lies within the Craven Basin (Hudson, 1933), also called the Bowland Basin by Ramsbottom (1974), Gawthorpe (1987) and Lee (1988).

The basin occupied an asymmetric graben which tilted south (Gawthorpe, 1987; Riley 1990b), and which was bounded by slowly subsiding regions known as the Southern Lake District High (Grayson and Oldham, 1987) and the Askrigg Block (Hudson, 1938) to the north, and the Central Lancashire High (Miller and Grayson, 1982) to the south.

Deposition started possibly in late Devonian times, and certainly by the early Dinantian (Courceyan). The nature of

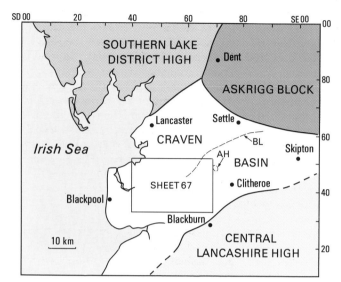

Figure 3 Dinantian palaeogeography (AH, Ashnott High; BL, Bowland Line).

the thick, concealed succession, suggested by seismic evidence (Chapter six) to lie beneath the oldest strata exposed (Chatburn Limestone Group), can only be speculated upon. By analogy with other contemporary extensional basins, such as the Widmerpool Gulf in the East Midlands (Smith and Smith in Plant and Jones, 1989), it is probable that the concealed sequence comprises a basal conglomerate, followed by red beds of "Old Red Sandstone" facies, overlain by peritidal deposits, which may include evaporites. The age of the earliest marine deposits is unknown, but by late Courceyan times a thick marine sequence of shallow-water limestones and fine terrigenous clastics (Charsley, 1984), the Chatburn Limestone Group (late Courceyan – early Chadian), accumulated. Depth of water appears to have been relatively uniform across the basin, despite the rapid subsidence that must have occurred to accommodate these strata, which locally appear to be over 3 km thick. Toward the basin margins, a relatively thin sequence of red beds, peritidal limestones and evaporites accumulated (for example the Stockdale Farm Formation of Arthurton et al., 1988).

This situation continued until the deposition of the Worston Shale Group (Chadian–Asbian), when differential subsidence began to exert a more obvious effect in the form of facies and thickness variations. At the base of the group, during deposition of the Clitheroe Limestone Formation (early Chadian), the southward tilt of the graben floor increased, resulting in deposition of shallow-water bioclastic limestones (Thornton Limestone Member) in the north and deeper-water Waulsortian limestones (Coplow and Bellman Limestone members) in the south. These deeper-water environments then progressively encroached northward. The closing phases of the deposition of the Clitheroe Limestone

Formation are unrecorded due to a widespread period of submarine erosion. The erosion probably resulted from back-scarp retreat along active faults which began to break up the basin floor at this time.

Deposition resumed during late Chadian times with the Hodder Mudstone Formation (late Chadian–Holkerian), which marks a switch in the main supply of carbonate sediment from sources within the basin to sources external to it. A hemipelagic, dysaerobic (oxygen poor) depositional regime also became established and persisted for the remainder of the Dinantian. The general southward tilt of the basin floor had steepened and it had fractured into a series of minor grabens and horsts, giving a varied topography. At first, this topography controlled sediment type and distribution, and produced a markedly diachronous and variable base to the formation. The most conspicuous deposits are boulder beds (Limekiln Wood Limestone), derived from the underlying Clitheroe Limestone Formation, and crinoidal packstones resulting from the profuse production of crinoidal detritus on the highs within the basin. In the northern part of the basin the deposition of relatively shallow water limestones (Hetton Beck Limestone Member) was re-established briefly. However, by Arundian times hemipelagic deposition was widespread, intrabasinal carbonate production had ceased and there was generally slow accumulation of fine-grained mudstones and calcisiltites in deeper water.

The change in sediment supply within the basin, referred to above, corresponded to the onset of shallow-water limestone deposition in the surrounding stable areas, where carbonate ramps and eventually platforms developed as these areas were submerged by marine transgression. On the Southern Lake District High, which was transgressed during late Chadian and Arundian times, these deposits are the Martin Limestone, Red Hill Oolite and Dalton Beds (Rose and Dunham, 1977); and on the Askrigg Block, submerged during Arundian times, they are the Chapel House Limestone and the lower part of the Kilnsey Formation (Arthurton et al., 1988). The platforms became the main source areas for carbonate sediment supplied to the basin throughout the rest of the Dinantian. Sediment supply was variable in type and volume as relative sea level oscillated in response to eustatic and tectonic mechanisms. During periods of relatively low sea level the carbonate platforms became emergent, allowing terrigenous clastics, such as the Buckbanks Sandstone Member, to reach the Craven Basin; conversely, during periods of relatively high sea level, carbonate production was optimised and a significant amount of detrital carbonate was exported into the Craven Basin to form accumulations of limestone turbidites, such as the Rain Gill and Chaigley Limestone members.

Basin floor topography maintained a significant control on the distribution of these sediments within the basin, in particular affecting the distribution of turbidites; the thickest accumulations generally built up in depressions on the basin floor or adjacent to active faults at the basin margins. Slumping and gravity slides (Riley, 1981; Gawthorpe and Clemmey, 1985) also resulted from unstable accumulations of sediment on slopes within the basin.

During Holkerian times, periods of emergence of the surrounding platforms became more frequent, inhibiting the export of carbonate into the basin, but allowing renewed influx of a small amount of terrigenous sand (for example in

the Cow Ark Anticline). The basin became progressively starved of detrital carbonate, a process that culminated in a basinwide accumulation of hemipelagic cephalopod limestones, the Hodderense Limestone Formation.

During late Holkerian and Asbian times, limestone turbidite supply resumed with the deposition of the Pendleside Limestone Formation. The carbonate platform environments surrounding the basin were at their maximum extent and copious amounts of detrital carbonate were exported into the basin. Periodic emergence of these platforms caused interruptions in carbonate supply and the local presence of thin sandstone turbidites (p.29).

Deposition of the Bowland Shale Group (late Asbian to Pendleian) began with a marked diminution in carbonate supply to the basin and an increase in oxygen starvation of the basin floor, resulting in the accumulation of black, organic-rich shales. This event can be correlated by means of ammonoid (goniatite) biostratigraphy with the initiation and growth of fringing reefs, for example the "reef-knolls" of the Cracoe district, at the boundary between the Craven Basin and the surrounding carbonate platforms ("Transition Zone" of Arthurton et al., 1988). The reefs isolated the basin from the surrounding carbonate platforms, the water column became stratified and the oxygen depleted, and black shale was deposited. Another effect of the fringing reefs was to partially obstruct the import of fine detrital carbonate into the basin, so that there was only localised accumulation of limestone turbidites, such as the Park Style and Ravensholme limestone members. Oscillations in relative sea level led to erosion of the platform margins, as indicated by the presence of limestone clasts and reworked foraminifera in some of the turbidites. This effect was also accentuated by uplift of the southern margin of the Askrigg Block (Arthurton et al., 1988).

The first major influx of terrigenous sand, caused by the approach of coastal deltas from the north, occurred in mid-Brigantian times (Pendleside Sandstones Member). The progressive spread of the deltas across northern Britain continued into Namurian times.

CLASSIFICATION

Lithostratigraphy

A comprehensive review of lithostratigraphical schemes was given by Fewtrell and Smith (1980). The lithostratigraphy adopted for the present survey is that of Riley (1990b), apart from the definition of the Lower Bowland Shale Formation and constituent members, which are detailed below. A summary of the classification used in this account is given in Figure 4, which also shows the corresponding biostratigraphical subdivisions; generalised sections are given in Figure 5. Lithological description of limestones follows Dunham (1962), and for Waulsortian buildups the terminology of Lees and Miller (1985) is used.

Biostratigraphy

Historical accounts of biostratigraphical schemes applied to the Dinantian sequence in the Craven Basin were given by Earp et al. (1961) and Arthurton et al. (1988). Riley (1990b)

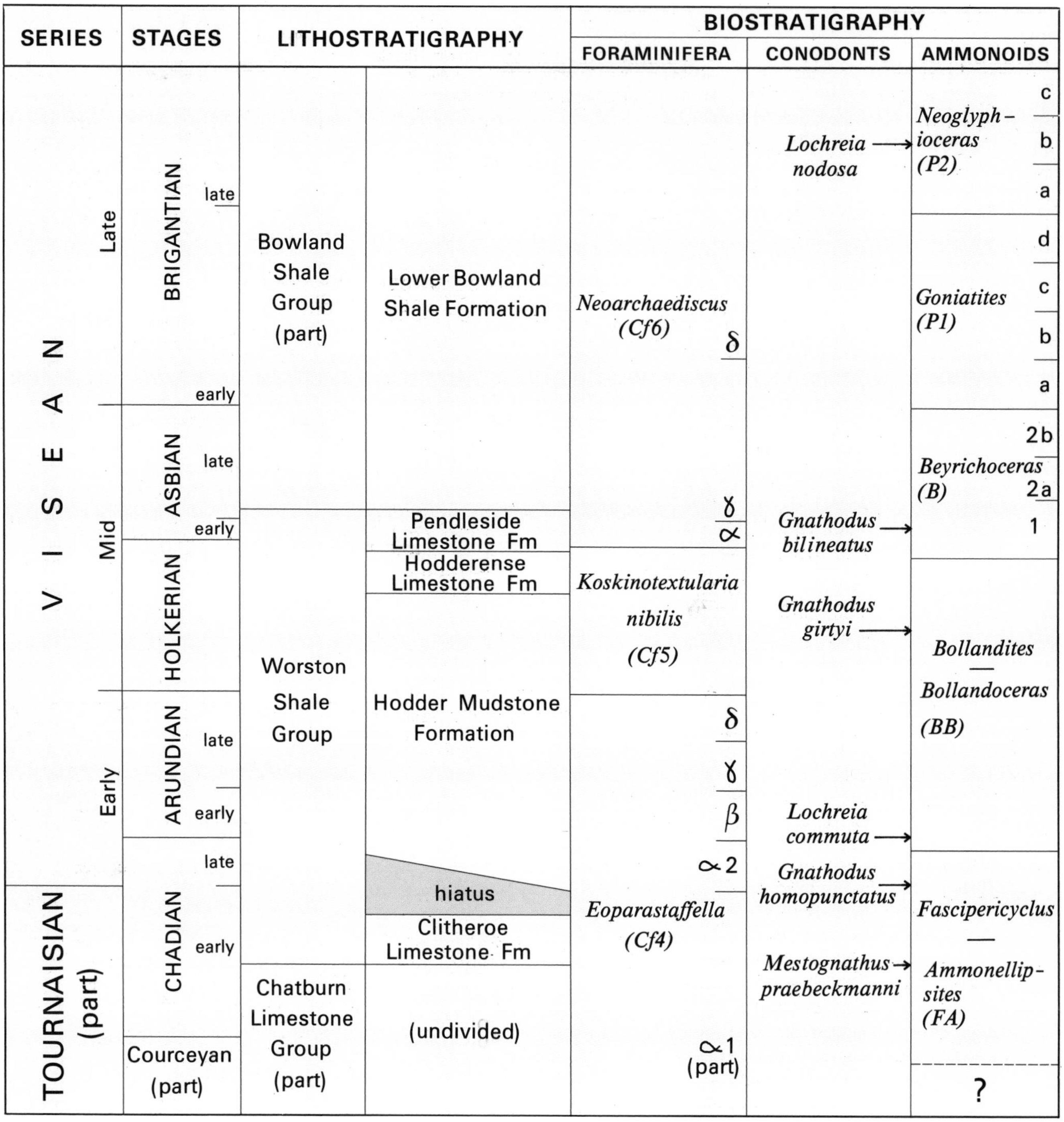

SERIES	STAGES			LITHOSTRATIGRAPHY		BIOSTRATIGRAPHY		
						FORAMINIFERA	CONODONTS	AMMONOIDS
V I S E A N	Late	BRIGANTIAN	late	Bowland Shale Group (part)	Lower Bowland Shale Formation	Neoarchaediscus (Cf6) δ	Lochreia nodosa →	Neoglyph- ioceras (P2) c / b / a
			early					Goniatites (P1) d / c / b / a
	Mid	ASBIAN	late			γ		Beyrichoceras (B) 2b / 2a
			early	Worston Shale Group	Pendleside Limestone Fm	α	Gnathodus bilineatus →	1
		HOLKERIAN			Hodderense Limestone Fm	Koskinotextularia nibilis (Cf5)	Gnathodus girtyi →	Bollandites — Bollandoceras (BB)
	Early	ARUNDIAN	late		Hodder Mudstone Formation	δ		
			early			γ		
						β	Lochreia commuta →	
		CHADIAN	late		hiatus	α 2	Gnathodus homopunctatus →	Fascipericyclus — Ammonellip- sites (FA)
			early		Clitheroe Limestone Fm	Eoparastaffella (Cf4)		
TOURNAISIAN (part)				Chatburn Limestone Group (part)	(undivided)		Mestognathus → praebeckmanni	
	Courceyan (part)					α 1 (part)		?

Figure 4   Dinantian stratigraphy and classification.

has summarised the ranges of principal guide fossils in the Worston Shale Group. A selection of Dinantian fossils from the present district is given on Plate 1.

Foraminifera and algae

Foraminifera are currently the most useful fossils for biostratigraphical subdivision of marine limestones in the middle and late Dinantian. Foraminiferal zonation in Western Europe has largely been based on the Belgian suc-cession and as yet there is no viable British-based zonal scheme. For this reason the zonation summarised in Paproth et al. (1983) is applied here, with minor modifications.

The oldest foraminiferal assemblages seen in the district occur in the Chatburn Limestone Group and belong to the basal subzone of the *Eoparastaffella* Biozone, Cf4α1. Significant components include *Bessiella*, *Brunsia*, *Dainella*, *Eotextularia*, *Florenella*, *Globoendothyra*, *Latiendothyra*, *Plectogyranopsis*, *Septabrunsiina*, *Spinobrunsiina* and *Spinoendothyra*, accompanied

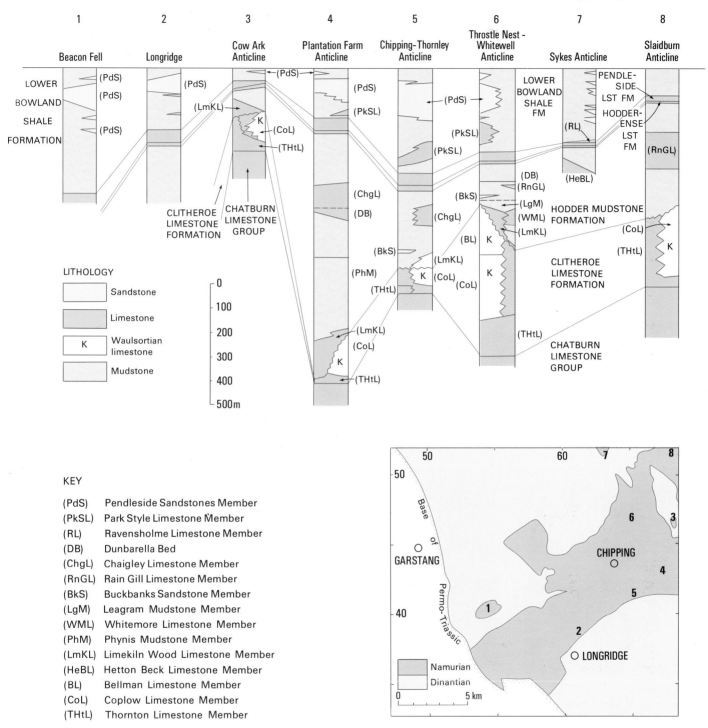

Figure 5 Generalised vertical sections through the Dinantian sequence.

by monolaminar *Koninckopora*, a dasyclad alga. *Mediocris* enters in the uppermost Chatburn Limestone Group. These assemblages persist into the Clitheroe Limestone Formation with some components becoming more abundant, particularly *Eotextularia, Florenella* and *Dainella*.

The earliest Cf4α2 Subzone assemblages occur in the basal beds of the Hodder Mudstone Formation. These are particularly well preserved and abundant in the Hetton Beck Limestone Member in the Sykes Anticline and in limestone

turbidites in the Cow Ark Anticline. Foraminifera of the previous subzone persist, with *Dainella, Eotextularia, Florenella* and *Mediocris* being conspicuous, but most significant is the first appearance of *Eoparastaffella* and the dasyclad alga *Koninckopora* with a bilaminar wall. Slightly higher in the sequence, within the lower part of the Hodder Mudstone Formation, the first archaediscids enter in the form of stunted *Uralodiscus* and *Glomodiscus*, together with *Planoarchaediscus* and *Viseidiscus*; these Cf4β Subzone assemblages are known

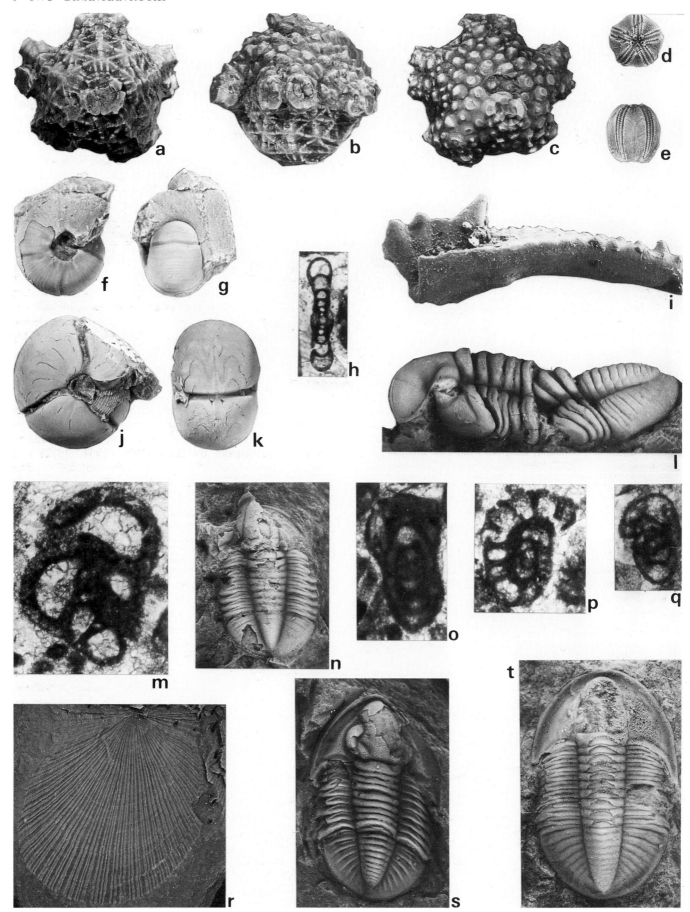

Plate 1   Selected Dinantian fossils (opposite)

a,b,c. *Actinocrinites* aff. *intermedius*, aboral, lateral and oral views of calyx respectively (Ro 6901) ×1.5. Hodder Mudstone Formation, Limekiln Wood Limestone Member, left bank of River Hodder, [6631 4330] Limekiln.

d,e. *Ellipticoblastus ellipticus*, oral and lateral views respectively (Ro 6824) ×1.5. Clitheroe Limestone Formation, Bellman Limestone Member, quarry [6443 4754] at disused calamine mine, Dinkling Green.

f,g. *Hammatocyclus* sp. nov. lateral and apertural views respectively (Ro 5127) ×2. Hodder Mudstone Formation, Chaigley Limestone Member, left bank of River Hodder [6830 4270] near Buck Hill, Bashall Eaves (same bed as Plate 7).

h. *Viseidiscus* sp., median axial section (MPK 6853) ×75. Hodder Mudstone Formation, turbidite channel [6348 4508] in left bank of Leagram Brook, Leagram.

i. *Mestognathus praebeckmanni*, oblique lateral view (MPK 6852) ×60. Clitheroe Limestone Formation, Coplow Limestone Member, Knoll Wood Quarry [6840 5000], near Newton.

j,k. *Goniatites granosus*, lateral and ventral views respectively (Ro 9847d) ×2. Lower Bowland Shale Formation, stream [6217 3935] above White Fold, Thornley.

l. *Bollandia persephone*, lateral view of distorted carapace (Ro 2731) ×3. Hodder Mudstone Formation, Phynis Mudstone Member, right bank of stream [6607 4699], Porter Wood, Whitewell.

m. *Eotextularia diversa*, axial section (MPK 6854) ×75. Hodder Mudstone Formation, Whitemore Limestone Member, stream [6472 4791], near Whitemore Knot, Dinkling Green.

n. *Latibole* sp. nov., dorsal view of carapace (Ro 4405) ×3. Hodder Mudstone Formation, Leagram Mudstone Member, abandoned mill race [6365 4450], Leagram Brook, Leagram.

o,q. *Eoparastaffella* spp., axial sections (MPK 6855 and 6856 respectively) ×75. Horizon and locality as Figure m.

p. *Pojarkovella* sp., saggital section (MPK 6857) ×75. Lower Bowland Shale Formation, Mill Brook, [6747 4485], near Cow Ark.

r. *Dunbarella* aff. *persimilis.*, left valve ×1.5. Hodder Mudstone Formation, Dunbarella Bed, within Chaigley Limestone Member, stream section [6803 4338], Paper Mill Wood.

s. *Weania feltrimensis*, dorsal view of carapace Ro 2779) ×3.5. Horizon and locality as Figure l.

t. *Phillibolina worsawensis*, dorsal view of carapace (Ro 4782) ×3.5. Clitheroe Limestone Formation, Bellman Limestone Member, Whitmore Knot [6455 4760], Dinkling Green.

to overlie the Buckbanks Sandstone Member in the Throstle Nest Anticline. Cf4γ Subzone assemblages with *Uralodiscus*, *Glomodiscus* and *Paraarchaediscus* are present in the Rain Gill and Chaigley Limestone members; the latter also includes Cf4δ Subzone assemblages containing large *Kasachstanodiscus*, previously referred to *Tubispirodiscus* (for example by Conil et al., 1980).

Assemblages referable to the *Koskinotextularia-nibelis* (Cf5) Zone are poorly represented due to the paucity of limestone turbidites in the upper part of the Hodder Mudstone Formation. Faunas in this district consist predominantly of stunted assemblages dominated by *Paraarchaediscus*, but *Pojarkovella nibelis* has been recovered from this interval in the Sykes Anticline in the adjacent Lancaster district.

The earliest *Neoarchaediscus* (Cf6) Zone assemblages occur in the basal beds of the Pendleside Limestone Formation. They belong to the Cf6a Subzone and include *Archaediscus* and *Paraarchaediscus* at the angulatus coiling stage, bilaminar palaeotextulariids, *Holkeria* and *Vissariotaxis*. The Cf6β Subzone does not appear to be recognisable in Britain, but profuse Cf6γ Subzone assemblages occur in the upper part of the Pendleside Limestone Formation and Lower Bowland Shale Formation. These include *Howchinia* and *Neoarchaediscus*. Reworked Cf5 Zone or Cf6α Subzone assemblages, with *Pojarkovella* sp., have been recognised in the Lower Bowland Shale Formation in the Cow Ark Anticline plate 1, fig. p. Elsewhere, for example in the Sykes Anticline, reworked Cf4γ–δ assemblages including *Eotextularia*, *Glomodiscus* and *Uralodiscus* occur at several horizons in the Cf5–Cf6 zonal interval.

Although ammonoid (goniatite) evidence indicates that the Lower Bowland Shale Formation should include foraminiferal faunas of the Cf6δ Subzone, no such foraminifera have been found, but the few samples collected indicate that Cf6γ Subzone foraminiferal assemblages persist up to the base of the Namurian. This anomaly is probably caused by facies effects and reworking.

Ammonoids

Ammonoids are an important faunal component, particularly in the hemipelagic sequences, where they provide a high level of biostratigraphical resolution. The Craven Basin is the type area for late Dinantian ammonoid zonation, developed principally by Bisat (1924, 1934, 1952) and Moore (1930a, 1936, 1939, 1946, 1950 and 1952). The district includes the famous Black Hall fauna, which was figured prominently by Phillips (1836). Dinantian ammonoid zonation was summarised in Ramsbottom and Saunders (1985), and Riley (1990a, 1991) has recently revised this zonation, partly on the basis of the sequence investigated during the present survey.

The oldest ammonoids occur in the Clitheroe Limestone Formation and probably lie in the *Fascipericyclus-Ammonellipsites* (FA) Zone, although the eponymous genera have not been found in this formation. Ammonoids include *Dzhaprakoceras* cf. *dzhaprakense* and *Rotopericyclus* cf. *rotuliformis*. Ammonoids first become abundant in the lower part of the Hodder Mudstone Formation, being most diverse in the Clitheroe Anticline where *Helicocyclus* and *Michiganites* are known. However, in the district late FA Zone faunas include *Ammonellipsites* ex gr. *kochi*, *Dzhaprakoceras* and *Merocanites* in the Phynis Mudstone Member, joined by *Eonomismoceras* in the Whitemore Limestone and Leagram Mudstone members. *Merocanites* persists throughout the rest of the Worston Shale Group.

The base of the succeeding *Bollandites-Bollandoceras* (BB) Zone lies in the middle part of the Hodder Mudstone Formation. *Bollandites*, *Bollandoceras* and *Hammatocyclus* have been recovered from the Chaigley Limestone Member. These faunas are very rare and have contributed greatly to our knowledge of the ammonoid record between the extinction of *Ammonellipsites* and the entry of *Beyrichoceras*. The uppermost beds of the Hodder Mudstone Formation mark the entry of *Dimorphoceras* and *Nomismoceras*, accompanied by *Bollandites*, *Bollandoceras* and *Merocanites*, which persist from earlier strata. These genera are represented in the Hodderense Limestone Formation by *Bollandites* sp. nov., *Bollandoceras hodderense*, *Dimorphoceras* sp. nov., *Merocanites* cf. *applanatus* and *Nomismoceras rotiforme*. Ammonoids are poorly known in the Pendleside Limestone Formation, although *Beyrichoceras* enters the basal beds to the south of the district, marking the base of the *Beyrichoceras* (B) Zone (Riley, 1990a).

The Lower Bowland Shale Formation contains profuse ammonoid assemblages ranging from the *Beyrichoceras* Zone (B1a Subzone), through to the top of the *Neoglyphioceras* Zone (P2c Subzone). The ammonoid sequence is that described by Earp et al. (1961) from the adjacent Clitheroe district, to which the reader is referred. The *Beyrichoceras* Zone extends into the lower part of the Park Style Limestone Member. At Black Hall Quarry (p.34) this member provided Phillips (1836) with the types of *Beyrichoceras obtusum, Beyrichoceratoides implicatum, B. stenolobus, Michiganites serpentinus,* and *Nomismoceras spirorbis*. It is probable that the types of *Dimorphoceras gilbertsoni* and *Nomismoceras rotiforme* also came from these strata or nearby sections. It appears that the upper part of the member lies in the P1c Subzone, but faunas are sparse. The main leaves of the Pendleside Sandstones Member occupy the same interval as that seen in the Clitheroe area, namely the P1c and P1d subzones.

Conodonts

A local conodont zonation scheme for the Craven Basin has been erected by Metcalfe (1981), and a British scheme has been proposed recently by Varker and Sevastopulo (1985). During the present survey the use of foraminifera was preferred; thus only sporadic conodont sampling was undertaken. In Figure 4, only the first entries of significant species are shown, rather than the zones and subzones. One of the most interesting discoveries has been that of *Mestognathus praebeckmanni* in the Coplow Limestone Member of the Clitheroe Limestone Formation at Knoll Wood (p.15, Plate 1i); this supports the foraminiferal correlation with the basal subzone (Cf4a1) of the *Eoparastaffella* Biozone (Cf4) for this formation.

Conodonts from the lower part of the Hodder Mudstone Formation have been recovered only from the Whitemore Limestone at Whitemore Knot (p.23); they correlate with the *Gnathodus pseudosemiglaber* Subzone of Metcalfe (1981). Higher strata exposed at Ravenscar Plantation contain abundant and well preserved conodonts of the *Gnathodus homopunctatus* Zone of Metcalfe (1981). Conodonts of this zone, with *Polygnathus bischoffi*, have also been found in the Chaigley Limestone Member.

The base of the *Gnathodus girtyi* Subzone was not located during the present survey; however, Riley (1990b) records the eponymous taxon from the upper part of the Hodder Mudstone Formation in the Clitheroe area. The succeeding *Gnathodus bilineatus* Zone enters in the upper part of the Pendleside Limestone Formation and extends into the Lower Bowland Shale Formation. Strata above the Park Style Limestone Member were not sampled for conodonts, but it is presumed that the base of the *Lochriea nodosa* Zone of Metcalfe (1981) lies within the P2b Ammonoid Subzone.

Corals and brachiopods

Before the use of microfossils, coral/brachiopod zonation schemes were traditionally employed. Much of the Craven Basin sequence lacks suitable facies, however, and stratigraphically significant coral/brachiopod faunas are rare. Full accounts of these assemblages were given by Parkinson (1926, 1935, 1936) and Earp et al. (1961).

Trilobites

The biostratigraphical potential of Dinantian trilobites is becoming realised and a full regional study within the Worston Shale Group was undertaken by Riley (1982, 1990b). In the Clitheroe Limestone Formation there are several new undescribed taxa of *Bollandia* and *Winterbergia*, which allow differentiation between the Coplow Limestone and Bellman Limestone members, facilitated by *Phillipsia gemmulifera*, which is found in the former, and *Phillibolina worsawensis*, in the latter.

The basal beds of the Hodder Mudstone Formation are rich in trilobites. The Phynis Mudstone contains *Weania feltrimensis*, known also from a similar horizon in the Dublin Basin of the Irish Republic, and *Bollandia persephone*. The succeeding Whitemore Limestone Member is dominated by *Weania gitarraeformis*, known also from the Milldale Limestones of South Staffordshire (Riley *in* Chisholm et al., 1988) and the Genicera Formation of Cantabria, Spain (Gandl, 1973). *Phillibole nitidus* and *Reediella stubblefieldi* also occur in this member; the former is best known from the famous Winterberg fauna of West Germany (G Hahn, 1966). The Leagram Mudstone Member contains a new, undescribed genus and *Latibole* sp. nov.. *Cummingella* sp. nov. is abundant in the Chaigley Limestone Member. The upper part of the Hodder Mudstone Formation yields a further new species of *Latibole*, which ranges into the lower part of the Pendleside Limestone Formation. Trilobites are abundant in the latter and lower part of the Lower Bowland Shale Formation, but there has been no detailed study. Faunas in the *Beyrichoceras* Ammonoid Zone of the Lower Bowland Shale Formation complement those known from West Germany, in the Cu IIIa Subzone of Nicolaus (1963), as described by G Hahn (1966), Hahn and Hahn (1971), Hahn et al. (1972) and Brauckmann (1973).

Echinoderms

Local correlation within the Clitheroe Limestone Formation and the lower parts of the Hodder Mudstone Formation can be achieved using blastoids. *Ellipticoblastus ellipticus* is common in the Bellman Limestone Member, but rare, if present at all, in the Coplow Limestone Member, the only record (unconfirmed) being that of Parkinson (1936) from Knoll Wood. *Mesoblastus* is known only from the Limekiln Wood Limestone Member.

Chronostratigraphy

Chronostratigraphical subdivision of the British Dinantian was proposed by George et al. (1976). Subsequently, various authors, including George (1978), Simpson and Kalvoda (1987), Leeder (1988), Davies et al. (1989) and Riley (1990b), have found several aspects of the scheme to be unsatisfactory. Most important is the lack of biostratigraphical identity to the basal boundaries of many of the stages, thus rendering precise correlation difficult. Stage and series correlations in this account therefore follow those suggested by Riley (1990b).

CHATBURN LIMESTONE GROUP (COURCEYAN – EARLY CHADIAN)

The Chatburn Limestone Group (Earp et al., 1961) comprises the oldest Dinantian rocks cropping out or proved in boreholes in the district. Thick Dinantian sequences of unknown facies are thought to lie at greater depth locally, notably south of the Thornley Anticline (p.80). The group includes several informal lithostratigraphic subdivisions used by Earp et al. (1961) during the survey of the Clitheroe district (Sheet 68). These units, which are not defined by stratotypes, comprise successively the Gisburn Coates Beds, Horrocksford Beds, Bankfield East Beds and Bold Venture Beds. Strata equivalent to the Bold Venture and possibly Bankfield East Beds appear to be represented in the present district, but no distinction has been made here.

Limestones of the Chatburn Limestone Group, which crop out in the core of the Slaidburn Anticline, were named the Slaidburn Limestone by Parkinson (1926, 1936) and assigned to his Pendleside Series. Fewtrell and Smith (1980) regarded the group as a formation, and defined a stratotype in Horrocksford Quarries at Chatburn in the adjacent Clitheroe district.

The group is widespread across the Craven Basin, and in the east, around Skipton and Broughton, it has been termed the Haw Bank Limestone (Hudson and Mitchell, 1937). In the Garstang district the group is conformably overlain by the Worston Shale Group. Outcrop is restricted to the cores of the Slaidburn, Throstle Nest and Whitewell anticlines; the group is also known in boreholes situated on the eastern closure of the Cow Ark Anticline.

In this district the limestone lithologies are generally monotonous and are dominated by mainly thin- to medium-bedded (20–70 cm thick), fine-grained, dark grey argillaceous packstones and subordinate dark, shaly, micromicaceous, calcareous mudstones and siltstones. Bedding surfaces are mostly planar. Bioturbation is extensive and has obscured much of the primary sedimentary fabric of the packstones. Principal grains are foraminifera, spicules (derived from sponges), brachiopods, ostracods and debris from fenestellid bryozoa and echinoderms. Algal oncolites are common in some beds. Macrofauna is only rarely abundant, but includes solitary corals, *Syringopora*, chonetoids, productoids, ribbed and smooth spiriferoids, large bellerophontid and euomphalid gastropods, rostraconchs and trilobites. Dolomite and chert replacement may occur and grain boundaries are commonly stylolitic, particularly in shaly beds.

The general depositional environment appears to have been a shallow marine carbonate shelf that received a relatively copious supply of fine mud and silt-grade terrigenous clastics. The substrate was well oxygenated, favouring deposit-feeding organisms and a specialised shelly benthos tolerant of relatively rapid sediment accumulation. The presence of oncolitic algae suggests that the sea, although probably turbid, was sufficiently shallow to allow light to reach the substrate.

Details

Slaidburn Anticline

The limestones form the core of the prominent south-west-trending feature of Pain Hill Moor and Moor End. Numerous exposures occur in small disused quarries and hillside screes. It is estimated that around 250 m of the upper part of the group is represented at outcrop. An interesting 2.9 m section is exposed in a disused quarry [6822 5103] west of Gamble Hole Farm. The basal 1.75 m comprises fine-grained packstones capped by a *Syringopora* colony. An erosive-based, poorly sorted, shelly packstone occupies the top 0.3 m of the section, resting upon a cross-bedded, fine-grained packstone. Good exposure also occurs in a disused quarry [6790 5083] 550 m north of Boarsden, and in steeply dipping sections exposed in disused pits and quarries [6825 5163] north-east of Bull Lane.

Whitewell Anticline

Exposures are limited to the Whitewell gorge. Several occur in both banks of the River Hodder [6571 4663 to 6536 4638], where the river runs along an anticlinal core. Good, but poorly accessible exposures also occur on the left bank [6547 4514; 6555 4525] in a valley near Ing Wood; about 100 m of the upper part of the group can be seen, showing typical lithologies of well-bedded, dark grey, fine-grained packstones and calcareous mudstones. Several exposures above the floodplain [6518 4558 to 6516 4570] yielded rich foraminiferal assemblages indicative of the Cf4α1 Subzone of the *Eoparastaffella* Foraminiferal Biozone.

Throstle Nest Anticline

Greystoneley Brook [6451 4469 to 6466 4494] affords the only exposure (Plate 2). Here, 73 m of Chatburn Limestone occur, with incomplete exposure in the top 48.5 m. The beds lie near the top of the group and display typical lithologies in the basal 18 m; however, the upper 55 m are transitional in character with the overlying Thornton Limestone Member.

Cow Ark Anticline

The group was proved in BGS Cow Ark boreholes SD 64 NE/7 and 10 (Figure 6). The contact with the overlying Worston Shale Group is variable, being represented by a thick mudstone interval in SD 64 NE/7, but by a more gradational passage into shaly, pale bioclastic packstones in borehole SD 64 NE/10. Lithologies were typical in both boreholes. Repeated faulting prevented calculation of accurate thicknesses, but borehole SD 64 NE/7 penetrated at least the top 18 m of the group.

WORSTON SHALE GROUP (CHADIAN – ASBIAN)

The Worston Shale Group comprises the Hodder Mudstone Formation, which dominates the sequence, and three thin and variable limestone formations, the Clitheroe Limestone, Hodderense Limestone and Pendleside Limestone formations.

Clitheroe Limestone Formation (early Chadian)

The term Clitheroe Limestone was applied by Parkinson (1926, 1935, 1936) to the limestones in the core of the Clitheroe Anticline, in the adjacent Clitheroe district. He subdivided the Clitheroe Limestone into the Chatburn

Plate 2   Chatburn Limestone in the gorge of the Greystoneley Brook near Greystoneley [6452 4470]. The gorge provides a good section in well-bedded, dark grey, fetid limestones with shaly partings and undulating bedding surfaces (A14758).

Limestone at the base, followed by the Coplow Knoll Series and Salthill Knoll Series above. Earp et al. (1961) retained the Chatburn Limestone, but included the upper two subdivisions of Parkinson's Clitheroe Limestone in their Worston Shale Group. Miller and Grayson (1972) applied the term Clitheroe Limestone Complex to these higher strata, whereas Arthurton et al. (1988) referred them to the Thornton Limestone of Hudson (1944). Fewtrell and Smith (1980) formally defined the base of the Clitheroe Limestone Formation at the junction with the Chatburn Limestone Group, but did not take into account the presence of a major erosion surface near the top of their formation. Riley (1990b) followed Fewtrell and Smith's (1980) definition of the base, but lowered the top of the formation to correspond with the erosion surface and defined several members within the formation. Riley's nomenclature is employed here.

The Clitheroe Limestone Formation is more variable both in facies and thickness than the underlying Chatburn Limestone Group, ranging from 100 m to around 1000 m thick. Generalised vertical sections are given in Figure 5. Three basic lithological assemblages occur. The most widespread consists of wavy-bedded, cherty, predominantly fine- to medium-grained, dark grey, argillaceous packstones and wackestones with scattered *Syringopora* colonies, caninioid corals and productoid/chonetoid/spiriferoid coquinas, together comprising the Thornton Limestone Member. These are interbedded with variable amounts of dark grey, commonly micromicacous, calcareous mudstones. The Thornton Limestone Member is best developed in the Slaidburn Anticline where it forms strong features, particularly on the northern limb. The junction with the underlying Chatburn Limestone Group is not exposed in the district, but is seen just beyond its eastern boundary at a disused quarry [6900 5256] north-east of New Biggin, in the adjacent Settle district (Arthurton et al., 1988). The junction has also been proved in boreholes in the eastern part of the Cow Ark Anticline.

Locally, the upper part of the Thornton Limestone Member passes laterally into a second major lithological assemblage which consists of very pure, biohermal or "reef" limestones of Waulsortian facies (see below). Two levels, the Coplow Limestone Member and the Bellman Limestone

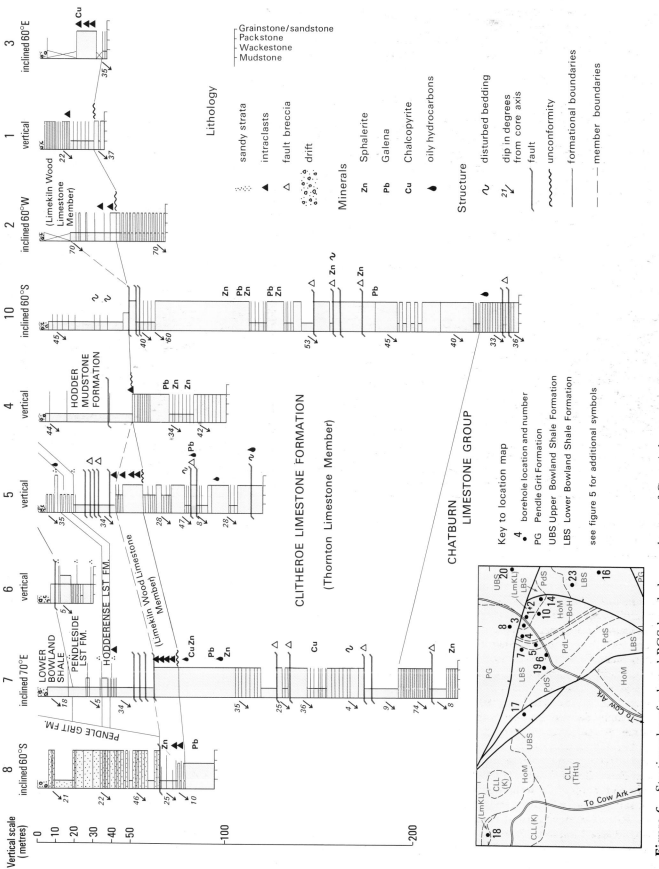

Figure 6 Stratigraphy of selected BGS boreholes north-east of Cow Ark.

Borehole numbers shown at the top of each column and on the location map are those of the BGS 1:10 000 Record System in which they are preceded by SD64NE. Other details are given in Appendix 1.

Member, are present, the latter being restricted to the northern limb of the Whitewell Anticline. The Coplow Limestone Member corresponds to Parkinson's (1926) Coplow Knoll Series, but the Bellman Limestone includes only the lower part of his Salthill Knoll Series, below Parkinson's (1936, p.302) "faunal break". These two members have not been distinguished on the geological maps, where they have both been given the symbol "K"; nevertheless, the distinction is made here to allow comparison with coeval sequences elsewhere in the Craven Basin. These "reef" limestones show dramatic thickness changes. They occur as discrete lenticular bodies of pale grey, poorly bedded packstones and wackestones, often forming steep-sided topographical features (Plate 3) from which is derived the traditional term "knoll reef".

The term Waulsortian was first applied during the last century to biohermal limestones and associated facies straddling the Tournaisian–Viséan boundary in the type area of the Dinantian in Belgium (Lees, 1988). This essentially lithostratigraphical usage was subsequently adapted and the term was used in a facies sense to describe similar limestones elsewhere (Lees, 1964). Recently, Lees et al. (1985), Lees and Miller (1985) and Miller (1986) have given a detailed description of these limestone facies and have proposed mechanisms for their origin and subsequent diagenesis. They have recognised a sequence of depositional phases, A, B,C and D, each based on a distinctive assemblage of petrological characteristics and bioclastic components that reflects the bathymetric changes brought about by the growth of the bioherms. Phases A to C represent a shallowing sequence of deposition in water too deep for light to penetrate, i.e. the aphotic zone below 220 m water depth (by comparison with modern equatorial waters); phase D represents deposition within the photic zone, between around 90 m and 220 m depth. Such detailed analysis of the Waulsortian was not attempted during the resurvey. However, Lees and Miller (1985) and Miller (1986) have given petrographic descriptions from various localities in the district and have recognised that the Waulsortian limestones here fall within phases B to D. The fact that algal structures are rare, and that the Waulsortian facies can develop in an aphotic environment, suggests that bacteria played a central role in their accumulation, possibly in the form of a gelatinous layer which initially covered and stabilised the substrate.

Several large-scale diagenetic and sedimentological features are regularly present. "Sheet-spars", which consist of layers of coarsely crystalline calcite-spar cement, are disposed parallel to the bedding. Since the host sediment boundaries are crudely refittable, such structures are considered to result from volume changes during early diagenesis. "Stromatactis" (see below) refers to lenticular-shaped cavities filled with calcite-spar cement and laminated dolomite-siltstone (dolosilt). Originally thought to be organic structures, their origin is problematical, probably involving a variety of depositional and diagenetic processes. "Fissures" comprise variously orientated, commonly vertical to subvertical fractures infilled with a variety of sediments and cements, including laminated dolosilts, calcite-spar cements, crinoidal packstones, grainstones and mudstones. Fissures reflect brittle fracturing of the Waulsor-

tian at a later diagenetic stage than the emplacement of sheet-spars. Infilling dolosilts accumulated predominantly during periods of vadose flow through the network of interconnected fissures by diagenetic fluids, whereas infilling bioclastic limestone and terriginous mud was a consequence of the fissures being open to the sea floor. The latter were commonly filled with sediments from the overlying Hodder Mudstone Formation, producing Neptunian dykes. Histories which involve progressive tilting of the Waulsortian limestones during diagenesis can be inferred from the geopetal characteristics of sediments and cements within the fissures and Stromatactis cavities.

At the base of the Waulsortian limestones and in places within or immediately lateral to them, there are thin beds comprising lithoclasts of Waulsortian limestone, usually associated with chloritic mudstone. In transverse section the diffuse and often deformed lithoclasts have a "clotted" texture. These intraformational breccias represent syndepositional disturbance during the early phases of Waulsortian bioherm formation and are termed "toe facies".

Surrounding the Waulsortian limestones of the Coplow and Bellman Limestone members is the third major lithological assemblage, which is particularly well developed in the Whitewell and Cow Ark anticlines. This assemblage, which includes "flank" and "interbank beds" (see below), has no named members. It comprises pale grey, medium- to coarse-grained crinoidal packstones and grainstones. In contrast to the Waulsortian limestones, these are generally thin-bedded, very crinoidal and commonly flaggy. They are also much cleaner washed, more crinoidal and paler than the limestone of the Thornton Limestone Member. Mudstone, if present, is usually limited to thin crinoidal partings. Dips in these beds commonly steepen close to the Waulsortian limestones, a combined effect of differential compaction and original depositional dips (palaeoslopes). Palaeoslopes of this kind can also be inferred from depositional fabrics within the Waulsortian limestones. Such dips indicate that many Waulsortian bioherms had considerable relief. Concentric strike patterns may also occur around the Waulsortian limestones in association with the steepening dips.

Miller and Grayson (1972) introduced a terminology which is useful for general description in the field, in which the Waulsortian limestones at the core of a bioherm are termed "bank beds". Each bank may consist of several stacked and interlocking mounds of Waulsortian limestone. "Flank beds" immediately surround the Waulsortian limestones and comprise thick-bedded to flaggy, crinoidal packstones and grainstones, commonly with depositional dips. The flank beds pass laterally into the more peripheral "interbank beds", comprising well-bedded crinoidal packstones with thin mudstone partings.

Details

Slaidburn Anticline

Only the south-western part of anticline lies within the district. The Thornton Limestone Member is well developed and forms prominent topographical features. On the north-western limb, small outcrops [6755 5058] east-south-east of Moor End lie less than 10 m above the base of the member and comprise wavy-bedded wackestones and packstones with chert nodules. The best section of the Thornton Limestone is discontinuously exposed in Red Syke [6716 5089 to 6721 5078], near Moor End, where 24.5 m of strata in the upper part of the member are overlain unconformably by the Hodder Mudstone Formation. Foraminifera obtained from here are consistent with a Cf4α1 Subzone correlation. Numerous small exposures in the middle and upper parts of the member occur on the hillside east and south-east of Rough Syke; they comprise shaly, cherty, wavy-bedded, crinoidal wackestones and packstones. On the south-eastern limb of the anticline there are only a few small exposures near the base of the member in Pot Wood [6785 5049] and Stubbings Wood [6809 5050], near Boarsden.

Waulsortian limestones of the Coplow Limestone Member form the steep-sided topographical features of Sugar Loaf (or Knot) [6708 5072] (Plate 3) and Knoll Wood [6840 5000], south-east of Boarsden, both of which have been quarried. At Sugar Loaf some 10 m of massive Waulsortian limestone are exposed; they are attributed to phases B,C and D (Lees and Miller, 1985; Miller, 1986). This outcrop lies on strike from the previously mentioned section of Thornton Limestone exposed 100 m to the north-east in Rough Syke, indicating that there must be a rapid lateral facies change. At Knoll Wood, about 15 m of massive Waulsortian limestone of phase D type (Lees and Miller 1985; Miller 1986) is well exposed in the quarry [6841 5005] where several fissures infilled with pale grey mudstone are present in the west face. Flank facies packstones and grainstones exposed at the bridge [6800 4987] near Knowlemere Manor are possibly contemporaneous with the Waulsortian limestone of Knoll Wood. Further exposure of Waulsortian limestone occurs upstream, 300 m to the south-west of Knoll Wood, in both banks of the River Hodder [6843 4964], forming a natural weir across the river ("Weir Knoll" of Earp et al., 1961). Here, the overlying Hodder Mudstone Formation is also seen.

The trilobite *Phillipsia gemmulifera* was found in the Waulsortian limestones at Sugar Loaf, at Knoll Wood and in the exposures at the weir, confirming correlation with the Coplow Limestone Member. The last locality is likely to have yielded a large fragment of the ammonoid *Rotopericylus* cf. *rotuliformis*, donated to the Geological Survey during the last century and labelled

Plate 3 A knoll-reef in the Clitheroe Limestone Formation, known as Sugar Loaf, situated in the Hodder Valley [670 507], near Dunsop Bridge. Though partially removed by quarrying, Sugar Loaf still preserves the typical knoll-like form of the Waulsortian buildups, which form important features of landscape in this part of the district. The Bowland Fells, formed from Pendle Grit, are in the background. They include the whaleback summits of Totridge, on the left, and Staple Oak Fell, in the right, separated by the valley of the Langden Brook (A14788).

"Knowlmere". During the present survey the conodont *Mestognathus praebeckmanni* (Plate 1i), characteristic of the Cf4α1 Subzone, was recovered from Knoll Wood.

Whitewell and Throstle Nest anticlines

On the northern limb of the Whitewell Anticline, the Thornton Limestone Member is discontinuously exposed in the deeply incised valley of Greystoneley Brook, below Greystoneley. The lowest beds, which occur [6442 4527] south-east of Greystoneley, comprise a 21.6 m-thick sequence of dark grey, argillaceous, fine-grained packstones and wackestones, and pale to medium grey, medium-grained packstones. Individual beds average 0.2 m in thickness and are bounded by calcareous, silty, micromicaceous mudstones. Bioturbation is common and the mudstones, in particular, are rich in macrofossils, especially crinoids and echinoids. Foraminifera indicating the Cf4α1 Subzone are abundant and include *Eblanaia michoti, Eotextularia diversa, Florenella* sp., *Globoendothyra* sp., *Mediocris* sp., *Priscella* sp. and *Pseudolituotubella* cf. *multicamerata*. The algae *Girvanella* sp., monolaminar *Koninckopora* and *Salebra sibirica* also occur. Small exposures of the upper part of the Thornton Limestone are seen [6412 4592] 400 m east of Lickhurst and [6441 4539] near Greystoneley. They consist predominantly of fine-grained packstones and wackestones, with a few beds of fine- to medium-grained packstones and grainstones. Thin nodular lenses of chert occur at some levels and some beds are irregularly dolomitised. Beds average 0.3 m in thickness, and have undulatory boundaries which may be poorly defined. Irregular "wavy" lamination is visible within some beds.

Beds transitional in character between Waulsortian limestone and the Thornton Limestone are exposed in a deeply incised gully 100 m east of Fair Oak Farm [6498 4602], where 4.2 m of irregularly bedded, pale grey, fine- to medium-grained, crinoidal packstones and grainstones are seen.

On the southern limb of the Whitewell Anticline, the Thornton Limestone is poorly developed due to a complex lateral facies change into Waulsortian limestone and related flank and interbank facies, which are particularly extensive here. No attempt has, therefore, been made to differentiate the member from the parent formation on the map in this area. A disused roadside quarry [6570 4660] in the Whitewell Gorge exposes about 10 m of interbank and flank facies within the basal 20 m of the formation. These strata consist of coarse-grained, well-bedded, pale grey, crinoidal packstones and grainstones and thin mudstones.

On the northern limb, similar strata low in the formation form a bench feature on the opposite side of the river gorge on the south-eastern flank of New Laund Hill [6555 4673], from which there appears to be a south-westward passage into Waulsortian limestone of the Coplow Limestone Member. There are numerous scree and crag exposures of Waulsortian limestone and associated facies on New Laund Hill. The trilobite *Phillipsia gemmulifera* was obtained from a small disused quarry [6488 4705] on the eastern flank of the hill, confirming the correlation with the Coplow Limestone. Lees and Miller (1985) referred these limestones to phases B and C.

Further south-west, in the western core of the Throstle Nest Anticline, the Coplow Limestone forms Knot Hill, where a disused quarry [6394 44753] exposes several faces, up to 8 m high, of massive Waulsortian limestone, comprising wackestones with sheet-spars. Fissures, infilled with laminated dolosilt, are well preserved and pervade many of the faces. One fissure intersects the Waulsortian limestone fabrics and forms an apparent dip slope on the south-eastern margin of the hill [6392 4463], outside the main quarry. In the north-western corner of the quarry [6390 4475], flank facies, comprising medium- to coarse-grained, crinoidal and bryozoan packstones and grainstones, overlie the Waulsortian limestones. This exposure has yielded the trilobites *Bollandia* sp. *Phillipsia gemmulifera* and *Weania* sp., together with the ammonoid

Dzhaprakoceras cf. *djaprakense*, confirming correlation with the Coplow Limestone.

On the northern limb of the Whitewell Anticline, younger Waulsortian limestones, referable to the Bellman Limestone Member, form spectacular steep-sided topographical features between Lickhurst Farm and Lower Fence Wood, and there are numerous small exposures and screes; the best exposures occur in small disused quarries [6385 4620; 6452 4661; 6488 4705; 6443 4754]. At the last locality there is also an adit to a disused calamine mine. In the quarry [6452 4661] south of Tunstall Ing, a 5 m-thick sequence of Waulsortian limestone, comprising massive pale wackestone with sheet-spars and Stromatactis-type cavities, is bounded by regularly bedded, flank facies packstones and wackestones. Bedding dips steeply to the north-east and is discordant to the north-westwards tectonic dip, suggesting an element of depositional dip here. Sheet-spar fabrics, again dipping discordantly to the regional dip, are well exposed in the western face of the quarry, north of Tunstall Ing [6443 4754]. Here the Waulsortian limestones are slightly dolomitised, resulting in differential weathering of the echinoderm fossils. Exceptionally well-preserved blastoids referable to *Ellipticoblastus ellipticus* (Plate 1d,e) are common at this locality. Thin veins of sphalerite and smithsonite (calamine) occur and were formerly worked in the adit. Small exposures of bank facies occur in woodland at the top of Whitemore Knot [6455 4760], north-north-east of Dinkling Green Farm, from which the trilobite *Phillibolina worsawensis* (Plate 1t) was obtained. These faunas support correlation with the Bellman Limestone Member.

Borehole SD 64 NW/2, 150 m north-west of Dinkling Green Farm, drilled Bellman Limestone, which is calculated to be 140 m thick when corrected for a tectonic dip of around 35°. Massive wackestones with sheet-spars predominate, with minor intercalations of packstones and grainstones referable to flank facies. Toe facies lithoclastic horizons, associated with greenish chloritic mudstones, also occur. Fissures and cavities showing spectacular geopetal fabrics (Plate 4a), variously filled with laminated micrite/dolosilt, dark grey mudstone, crinoidal packstone, calcite-spar and ferroan dolomite, intersect the borehole at several horizons. Oil bleeds and pyrobitumen are present in vughs in the upper 60 m of the sequence (Plate 16).

Cow Ark Anticline

As in the Whitewell Anticline, much of the Thornton Limestone Member passes laterally into Waulsortian limestones of the Coplow Limestone Member and associated flank and interbank facies; hence the Thornton Limestone has not been differentiated on the map. An exposure of typical Thornton Limestone is seen in a small disused quarry [6745 4651], 130 m north-east of Higher Park Gate Farm, where at least 1.8 m of shaly, wavy-bedded, dark grey, crinoidal packstones and wackestones are exposed, yielding calyces of the crinoid *Actinocrinites* and the trilobite *Phillipsia gemmulifera*. The Thornton Limestone Member was also penetrated in boreholes SD 64 NE/1,2,4,7,8,10,14 and 16 (Figures 6 and 7). The most com-

Plate 4 Slabbed core of a) Waulsortian limestone in the Bellman Limestone Member (early Chadian), showing sheet spar fabrics disposed at a high angle to the core axis, with later geopetals of ferroan dolosiltstone. Borehole SD 64 NW/2, 14.10 to 14.40 m depth.
and b) Limekiln Wood Limestone (late Chadian), showing breccia of Clitheroe Limestone clasts. Many clast contacts are stylolitised. Calcite spar-lined cavities are impregnated with metalliferous sulphide mineralisation. Borehole SD 64 NE/7, 51.29 to 51.60 m depth.

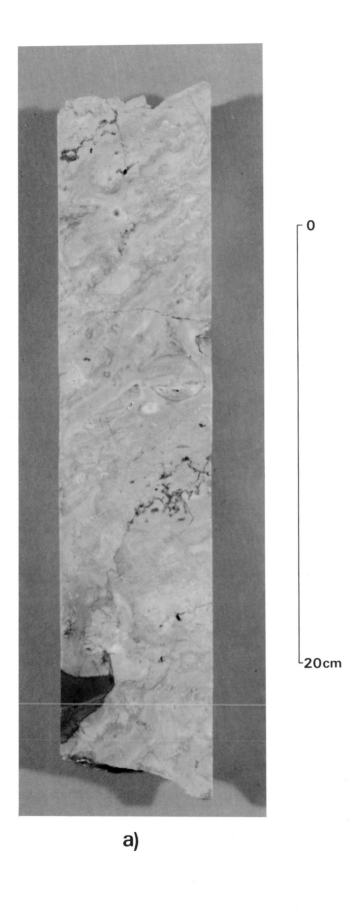

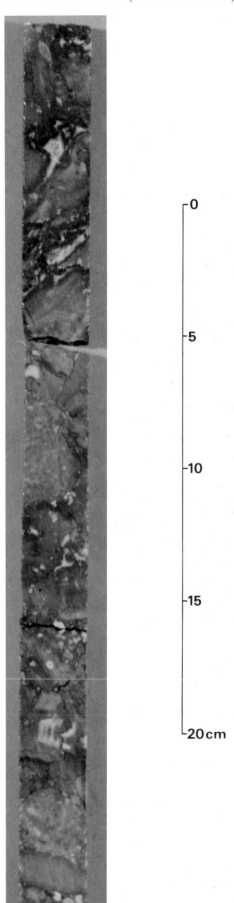

0

20cm

0

5

10

15

20cm

a)

b)

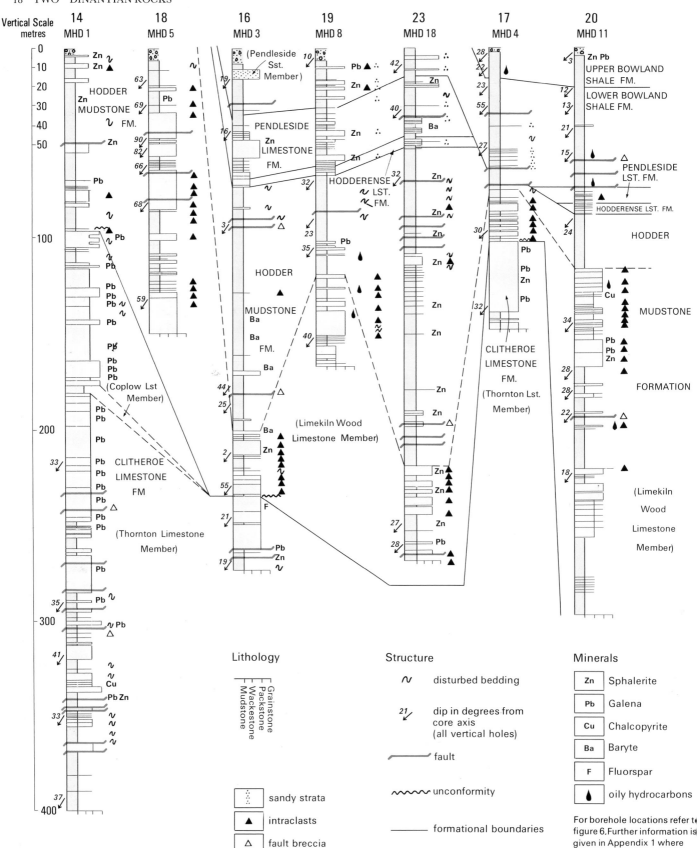

Figure 7   Stratigraphy of selected BP Minerals Boreholes (MHD series) north-east of Cow Ark (for location of boreholes see map attached to Figure 6).

Borehole numbers shown at the top of each column are those of the BGS 1:10 000 Record System in which they are preceded by SD64NE. Other details are given in Appendix 1.

plete sections occur in SD 64 NE/7 (Plate 4b) and 64 NE 10 including the junction with the underlying Chatburn Limestone Group. In all these boreholes north-east of Cow Ark, the Thornton Limestone Member is unconformably overlain by the Hodder Mudstone Formation, with the exception of SD 64 NE/14 where the Thornton Limestone is overlain by Waulsortian limestones of the Coplow Limestone Member. Foraminiferal assemblages obtained from the Thornton Limestone in these boreholes are typical of the Cf4α1 Subzone and include *Brunsia* sp., *Eblanaia michoti*, *Endothyra laxa*, *Eotextularia diversa*, *Florenella* sp., cf. *Globoendothyra* sp., *Palaeospiroplectammina mellina*, *Pseudolituotubella multicamerata*, *Septabrunsiina* sp., *Spinobrunsiina* sp., *Spinoendothyra* sp.; algae including *Girvanella* sp., monolaminar *Koninckopora* and *Salebra sibirica* also occur.

Waulsortian limestones of the Coplow Limestone Member, and associated flank and interbank facies, are well exposed on the northern limb of the Cow Ark Anticline. The vertical and lateral passage of interbank and flank facies into the overlying Waulsortian limestones is seen in a disused quarry at Hall Hill [6675 4660], east of Whitewell. This exposure, which was described by Black (1952, 1954), is the most impressive in the Craven Basin for displaying the contact between the Waulsortian limestones and underlying strata. The former fall within phase B, to possibly phase D, in Lees and Miller's (1985) scheme. The lower part of the quarry, around the old limekiln, shows about 4 m of flank and interbank facies comprising thin-bedded units of wavy-bedded, pale grey, medium- to coarse-grained, siliceous, crinoidal packstones and thin shales, interbedded with thicker beds of coarse-grained, pale grey, crinoidal packstone and grainstone with wavy laminae. Crinoid calyces are common and include *Actinocrinites* and *Gilbertsocrinus*. The trilobite *Phillipsia gemmulifera* is also common, confirming correlation with the Coplow Limestone Member. In the main part of the quarry, north of the limekiln, these facies pass into Waulsortian limestones comprising poorly bedded wackestones with sheet-spars. The contact between the Waulsortian and the underlying strata is markedly diachronous, being some 3 m higher at the western end of the quarry face than along strike at the eastern end. This relationship appears to be primarily depositional, although some soft sediment loading and minor shearing occurs along the contact, probably due to differential compaction. Toward the western end of the face, interbeds of toe facies comprise breccias of angular, platy, laminated, wackestone clasts set in a packstone/wackestone matrix. Toward the eastern end of the quarry, there is a series of large fissures obliquely intersecting the bedding, which are filled with fine- to coarse-grained crinoidal packstone and grainstone. Fractures along these fissures have given rise to pseudobedding.

Poorly bedded Waulsortian limestones of the Coplow Limestone Member, interbedded with crinoidal mudstones and thin, medium-grained packstones, occur in a large swallow hole at Ravenscar Plantation [6676 4695], 300 m to the north-east of the Hall Hill quarry. The contact with the overlying Hodder Mudstone Formation is also exposed, and further exposure of the contact is seen downstream from a cave sump in Porter Wood [6607 4699].

At a waterfall [6730 4539] immediately downstream from Cow Ark Bridge, a rich Cf4α1 Subzone foraminiferal assemblage was obtained from flank and interbank facies, including *Eotextularia diversa* and cf. *Latiendothyranopsis menneri*; also kamaenid algae. Higher in the sequence, small exposures of Waulsortian limestone referable to the Coplow Limestone Member occur immediately upstream from the road bridge, near an old limekiln.

Borehole SD 64 NE/14 (Figure 7) proved a thin development of Waulsortian limestone belonging to the Coplow Limestone Member, overlain by a thick sequence of flank facies packstones and grainstones, which persists to the top of the formation. The trilobite *Phillipsia gemmulifera* was found in these beds, confirming correlation with the Coplow Limestone.

Thornley Anticline

Because of thick drift, it is not known whether the Thornton Limestone crops out in the core of the anticline. Exposures of the Clitheroe Limestone Formation are limited to Waulsortian limestones and these are thought to belong to the Coplow Limestone Member. They occur at Knott Farm [614 403], Arbour Farm [619 407] and Leach House [628 413], and form topographical features which have a south-west trend.

Knott Farm lies on a low hill; exposure is seen immediately south-west of the farm [6140 4021] and comprises 2.2 m of massive, pale grey wackestones in bank facies.

At Arbour Farm the Waulsortian limestones have been largely quarried away, but they are still exposed in the now disused quarry. Much of the south and west faces of the quarry comprise limestones referable to the Hodder Mudstone Formation (p.24), but on the south-east face about 3.2 m of Waulsortian limestone is exposed, comprising massive, pale grey, micritic wackestones with sheet-spars. A sharp undulating contact separates this from the underlying bioclastic, flank facies packstones. A similar junction is present at an isolated exposure in the middle of the quarry, except that the sequence is reversed; here 1.5 m of Waulsortian limestone, with sheet-spar fabrics, is overlain by about 9.0 m of coarsely crinoidal packstones of flank facies. In the north-western face there is a breccia with angular clasts of Waulsortian limestone set in a micritic matrix, showing concentric lamination; some clasts show edges which are refittable with adjacent clasts. This quarry was described also by Parkinson (1935, pp.112–116).

At Leach House, Waulsortian limestones are seen in a disused quarry; they are best exposed in a 4 m-high face at its northern end [6284 4132]. The most prominent face, in the south-eastern part of the quarry, is broken and brecciated, probably by faulting, but no mappable displacement is evident.

Hodder Mudstone Formation (late Chadian – Holkerian)

This formation was first proposed by Riley (1990b). It corresponds closely to the Worston Shale Series of Parkinson (1926), the Worston Series of Parkinson (1935, 1936), the Worston Shales of Earp et al. (1961) and Arthurton et al. (1988), and the Salthill Cap Beds and Cephalopod Shales of Miller and Grayson (1972). Fewtrell and Smith's (1980) Worston Shale Formation excludes the lower part of the Hodder Mudstone Formation and extends higher to include the overlying Hodderense Limestone Formation ("*Bollandoceras hodderense*" beds).

In the district the Hodder Mudstone has a maximum thickness of 900 m around the Plantation Farm Anticline and a minimum of less than 200 m in the eastern closure of the Cow Ark Anticline and in the northern limbs of the Whitewell and Throstle Nest anticlines. Because of its argillaceous nature the formation generally gives rise to subdued topography, but interbedded sandstone and limestone members form local features.

The formation succeeds a widespread unconformity developed on the Clitheroe Limestone Formation, which locally had considerable relief. Consequently, in its lower part the Hodder Mudstone Formation is diachronous and heterolithic, with boulder beds and breccias comprising clasts derived from the Clitheroe Limestone Formation, interbedded with mudstones, limestone turbidites and, locally, debris flows. In the northern part of the district (Sykes Anticline) shelf limestones occur in the basal part. The middle and upper parts of the formation are less variable; hemipelagic mudstone and thin-bedded, laminated calcisiltites predominate, with local intercalations of sandstone and limestone turbidites. Soft sediment deformation, in the form of slumping and gravity slides, is common. The formation is succeeded conformably by hemipelagic cephalopod limestones of the Hodderense Limestone Formation.

The mudstones in the Hodder Mudstone are predominantly calcareous, pale to dark grey and blocky. Small-scale bioturbation is common, but it rarely obliterates depositional laminae. Finely micaceous and fissile mudstones also occur. The autochthonous macrofauna is generally limited to solitary corals, small chonetoids, smooth spiriferoids, pectinoid and nuculoid bivalves, gastropods, cephalopods, trilobites and fish debris, suggesting a dysaerobic environment, one in which oxygen levels at the substrate surface were depleted; plant debris may also occur. Only the mudstones near the base of the formation contain a more diverse assemblage, which includes additional echinoderms, brachiopods and bryozoans.

Limestone turbidites are locally common, especially towards the base. They include limestone clast breccias, debris flows and graded packstones. Some beds show erosive bases, sole markings, rip-up clasts and discontinuous wavy and convolute lamination and slumping; there are some composite beds also. In contrast to the mudstones, these turbiditic limestones commonly contain an allochthonous, fragmentary fauna including thick-shelled brachiopods and crinoids. Foraminifera and algae are also common, indicating derivation from shallow water environments.

Several members are recognised, and in the lower part of the formation some are markedly diachronous. The base of the formation is usually represented by the Limekiln Wood Limestone Member, comprising detrital crinoidal limestones, limestone breccias and boulder beds, interbedded with variable amounts of crinoidal mudstone. The member is equivalent to the Salthill Cap Beds of Miller and Grayson (1972), and is particularly well developed over the eroded Waulsortian limestones of the underlying Clitheroe Limestone Formation in the Whitewell and Thornley anticlines. The thickest development, over 125 m, is in the Plantation Farm Anticline (Figure 5). Elsewhere, for example on the northern limb of the Slaidburn Anticline, the Limekiln Wood Limestone is very thin and much of the lower part of the formation comprises the Phynis Mudstone Member, which is a distinct unit of blocky, dark grey, finely micaceous, calcareous mudstone. The member is equivalent to the Phynis Shales of Parkinson (1936). It has a general thickness of less than 50 m, but is over 350 m thick in the Plantation Farm Anticline. It is present also in the Whitewell and Slaidburn anticlines. Features are poor, and the member has therefore not been distinguished on the map. The Whitemore Limestone Member overlies the Phynis Mudstone and comprises pale grey, shaly wackestones and turbiditic, detrital limestones, interbedded with pale to dark grey, platy, calcareous mudstones. The member is feature forming and has a general thickness of 50 m . It is present in the Slaidburn and Whitewell anticlines, and on the northern limb of the Plantation Farm Anticline. The succeeding Leagram Mudstone Member, equivalent to the Prolecanites

compressus Beds of Parkinson (1926), contains platy and blocky, dark blue, pyritic mudstones, interbedded with dark, diffusely bedded, argillaceous wackestones. Crushed conches of the ammonoid *Merocanites* are common. This member generally shows poor features and hence is not distinguished on the map. It is present in the Whitewell and Slaidburn anticlines, where it has a general thickness of 50 m. The Phynis Mudstone, Whitemore Limestone and Leagram Mudstone members interleave laterally with the Limekiln Wood Limestone.

In the Sykes Anticline the lowest exposed beds of the Hodder Mudstone Formation belong to the Hetton Beck Limestone Member, which comprises well-bedded, argillaceous, fine- to coarse-grained packstones, with graded bedding and slumping common in the upper part. Autochthonous shelly faunas, including *Syringopora* colonies, are common. The member is at least 120 m thick; its base is not seen.

The Buckbanks Sandstone Member, in the middle part of the Hodder Mudstone (Figure 5), comprises fine- to medium-grained, calcareous, micaceous, quartzitic sandstone with scattered crinoid and brachiopod debris. When fresh it is dark grey and very hard, but on weathering it is ochreous and friable. The member is restricted to the northern limbs of the Thornley and Throstle Nest anticlines, with a maximum thickness of 11.5 m. Younger members, also within the middle part of the formation (Figure 5), include the Rain Gill Limestone Member, equivalent to the Middop Limestone of Earp et al. (1961), which comprises diffusely bedded, slumped, turbiditic packstones and wackestones and the slightly younger Chaigley Limestone Member, which is made up of well-bedded, turbiditic, detrital packstones interbedded with fissile and blocky mudstones and siltstones. The Rain Gill Limestone is present in the Slaidburn Anticline, where it forms good features and is up to about 148 m thick, and in the eastern part of the Whitewell Anticline. The Chaigley Limestone occurs in the Plantation Farm and Thornley anticlines, where it is up to 150 m thick and forms low features. The Dunbarella Bed lies in the lower part of the Chaigley Limestone Member in the Plantation Farm Anticline. This unit, which is usually less than 0.3 m thick, comprises black, fissile mudstone crowded with articulated valves of the bivalves *Dunbarella* and *Pteronites*. The bed also occurs in the Whitewell and Throstle Nest anticlines, allowing correlation (Figure 5) into areas where the Chaigley Limestone is absent.

Details

Cow Ark Anticline

There are a number of sections in the Cow Ark Anticline which expose the erosive contact with the underlying Clitheroe Limestone Formation. The basal beds of the Hodder Mudstone Formation onlap against a considerable topography developed on the underlying Clitheroe Limestone, and for this reason they are markedly diachronous. On the northern limb, the oldest beds exposed belong to the Limekiln Wood Limestone Member and occur in a stream section at Porter Wood [6607 4699], north-east of Whitewell, immediately downstream from a cave sump. They lie with sharp contact upon Waulsortian limestones of the Clitheroe Limestone (Coplow Limestone Member) and comprise 0.5 m of thin-bedded, very coarse-grained, crinoidal packstones and grainstones, overlain by 8 m of the Phynis Mudstone Member. This latter unit is well exposed in a 3 m-high meander cliff on the right bank, downstream from the sump, and consists of dark grey, micromicaceous, calcareous mudstone yielding a prolific macrofauna including solitary corals, cf. *Paraconularia* sp., fenestrate bryozoans, *Lingula* sp., *Orbiculoidea* sp., chonetoids, productoids, *Schizophoria resupinata*, smooth and ribbed spiriferoids, *Aviculopecten* sp., nuculoids, gastropods, orthocone and stroboceratid nautiloids, *Ammonellipsites* ex gr. *kochi*, *Merocanites* sp., *Bollandia persephone* (Plate 1l), *Brachymetopus* ex gr. *moelleri*, *Weania feltrimensis* (Plate 5s), *Cyathocrinus* sp., *Gilbertsocrinus* sp., cf. *Callograptus* sp. and fish debris. Discontinuous exposure further downstream includes a few erosive-based, graded beds of crinoidal packstone with scattered packstone intraclasts, suggesting that there is a lateral passage nearby into the Limekiln Wood Limestone Member. These graded beds yield the conodonts *Gnathodus pseudosemiglaber*, *Mestognathus beckmanni* and *Polygnathus bischoffi*, accompanied by sparse foraminifera including *Eblanaia* sp., *Eotextularia diversa* and the alga *Girvanella*. The conodonts indicate correlation with the *Gnathodus pseudosemiglaber* Subzone of Metcalfe (1981) and the ammonoid *Ammonellipsites* indicates correlation with the upper part of the *Fascipericyclus-Ammonellipsites* Zone of Riley (1991).

On the northern limb, an important exposure is seen [6677 4695] east of Whitewell, at Ravenscar Plantation, immediately upstream from a swallow hole. Here the basal beds of the formation consist of a thin veneer of Limekiln Wood Limestone, comprising thin-bedded, very coarse-grained crinoidal packstones, resting with a sharp erosive contact on the Waulsortian (Coplow Limestone Member) of the Clitheroe Limestone Formation. Upstream, a sequence of platy and fissile, dark grey to black, calcareous mudstones and thin laminated calcisiltites is discontinuously exposed up to a tributary intersection [6693 4696]. These beds are lithologically similar to strata seen elsewhere immediately overlying the Leagram Mudstone Member; they yield the trilobite *Liobole* sp., which first enters the local sequence in the Leagram Mudstone Member, implying that deposition upon the eroded Clitheroe Limestone Formation was initiated later than at Porter Wood. Upstream from the tributary junction, the dip steepens almost to vertical, and erosive-based, graded beds of fine-grained packstone are interbedded with the mudstones, forming a series of small waterfalls up to the boundary wall [6686 4694]. These beds contain a new trilobite taxon which has not been seen below the Leagram Mudstone Member, accompanied by crushed *Dzhaprakoceras* sp. and *Merocanites* sp.. Foraminifera indicative of the Cf4α2 Subzone are abundant in the packstones and include *Endothyra* sp., *Eoparastaffella simplex*, *Eotextularia diversa*, *Florenella* sp., lituotubellids, *Mediocris* sp., *Tetrataxis* sp. and *Valvulinella* sp.; algae including kamaeniids, monolaminar and bilaminar *Koninckopora* spp. and stacheiinids also occur. Conodonts are well preserved and prolific; they include *Gnathodus homopunctatus*, *G. pseudosemiglaber*, *Mestognathus beckmanni* and *Polygnathus bischoffi*. Further small, weathered exposures of dark grey mudstone and rubbly, pale grey wackestone, yielding *Bollandia persephone*, occur nearby in small streams [6694 4676; 6710 4679]. The western exposure shows a sharp undulating contact with the Coplow Limestone Member.

On the southern limb of the anticline, beds low in the Hodder Mudstone Formation are exposed along Hagg Clough [6731 4557 to 6773 4608], upstream from Cow Ark. The section is discontinuous and greatly disturbed by faulting. The Limekiln Wood Limestone is seen [6731 4557] upstream from Cow Ark road bridge and comprises well-bedded limestone breccias with angular clasts of Waulsortian limestone derived from the Clitheroe Limestone exposed nearby. Upstream, the beds become more shaly and breccias are succeeded by graded beds of coarse- to fine-grained crinoidal packstone. The highest strata, comprising cleaved, blocky, finely micaceous mudstones belonging to the Phynis Mudstone, are exposed further upstream [6773 4608].

North-east of Cow Ark, contacts between Hodder Mudstone and the Clitheroe Limestone were penetrated by boreholes SD 64 NE/1,2,4,5,7,8,10,14,16 and 17 (Figures 6 and 7). Apart from borehole SD 64 NE/14, where the contact is on flank and interbank facies overlying the Coplow Limestone Member, the highest proved beds of the Clitheroe Limestone all belong to the Thornton Limestone Member. In boreholes SD 64 NE/8 and 10 the contact is faulted; in the other boreholes contacts are sharp and unconformable. In all cases the basal beds of the Hodder Mudstone belong to the Limekiln Wood Limestone Member and comprise thin-bedded and composite beds of limestone breccia composed of angular clasts of Waulsortian limestone, commonly with stylolitised boundaries, set in a crinoidal packstone matrix (Plate 5). These limestones are interbedded with various proportions of finely micacous, dark grey to black, blocky, calcareous mudstone. Some limestones are erosive-based and show graded bedding. The Limekiln Wood Limestone was also proved in boreholes SD 64 NE/18 and 20. Due to faulting it is difficult to estimate the maximum thickness of the member, but it is in the order of 200 m. The Phynis Mudstone, Whitemore Limestone and Leagram Mudstone members were not recognised. It is not clear whether their absence is due to faulting, non-deposition or lateral facies change into the Limekiln Wood Limestone. Reworked Cf4α1 Subzone assemblages recovered from the Limekiln Wood Limestone in these boreholes include the foraminifera *Brunsia* sp., *Eblanaia michoti*, *Eotextularia diversa*, *Florenella* sp., cf. *Latiendothyranopsis* sp., *Lugtonia monilis*, *Mediocris* sp.; the algae *Girvanella* sp., monolaminar *Koninckopora* and stacheiinids also occur.

Higher beds in the formation, extending up to the succeeding Hodderense Limestone Formation, proved in boreholes SD 64 NE/5,7,16,19,20 and 23, have no formal member name. They comprise dark grey, subfissile to blocky, calcareous mudstone interbedded with various amounts of predominantly thin-bedded, fine-grained packstone, laminated calcisiltite and diffusely bedded, dark grey, argillaceous wackestone. Veins of fibrous calcite are common, together with listricated zones, tectonically disturbed bedding and faulting. The thinnest, apparently unfaulted sequence occurs in borehole SD 64 NE/20, where about 13 m of mudstone (when corrected for dip) is present between the top of the formation and the Limekiln Wood Limestone. The equivalent, thicker sequences in boreholes SD 64 NE/16 and 23 are heavily faulted, but an estimated thickness of around 140 m is suggested when corrected for dip.

Throstle Nest Anticline

The most complete exposure [6381 4435 to 6323 4514] of the Hodder Mudstone Formation is in the stream flowing east of Dairy Barn, known as Leagram Brook. The lowest beds exposed occur in the core of the anticline in a 5 m-high meander cliff on the east bank [6377 4448]; they comprise blocky, calcareous, dark mudstones and diffusely bedded, argillaceous wackestones which belong to the Leagram Mudstone Member. These beds are estimated to lie about 20 m above the Waulsortian limestones of the Clitheroe Limestone Formation exposed nearby at Knot Hill (p.16). Downstream from the meander, dips steepen into the southern limb of the anticline and the beds become tightly folded, with numerous small wrench faults. Here, the upper part of the Leagram Mudstone and overlying strata, comprising subfissile and platy dark mudstones, laminated calcisiltites and thin sandy packstones with graded bedding and sole markings, are exposed [6377 4445]. Foraminifera from the packstones belong to the Cf4α2 Subzone and include *Brunsia* sp., *Endospiroplectammina* sp., cf. *Eoparastaffella* sp., *Mediocris* sp., *Plectogyranopsis* sp., *Pseudolituotubella* sp., *Septabrunsia* sp. and *Spinobrunsiina* sp., together with algae including kamaenids and monolaminar *Koninckopora* sp.. The highest beds [6381 4435] comprise ochreous weathering, hard, grey, fine- to medium-grained, calcareous sandstone of the Buckbanks Sandstone Member, interbedded with sandy packstones and thin mudstones. This exposure, which is the type locality (Riley 1990b), for the member, reveals up to 2.63 m of beds.

Upstream from the anticlinal core, dips are shallow and the stream meanders repeatedly cross the strike of the Leagram Mudstone, which is at least 40 m thick. Good exposure occurs in the stream and in an adjacent abandoned mill race on the right bank. Macrofauna obtained from these beds includes solitary rugose corals, chonetoids, *Crurithyris* sp., *Nudirostra papyracea*, *Aviculomya* sp. *Aviculopecten* sp., nuculoids, bellerophontids, orthocone and stroboceratid nautiloids, *Dzhaprakoceras* spp., *Eonomismoceras* sp., *Merocanites* sp., *Latibole* sp. nov. (Plate 1n), *Liobole castroi*, a new, undescribed trilobite and crinoid debris. This is the first entry of the trilobite *Liobole* in the local sequence and the ammonoids indicate a level in the upper part of the *Fascipericyclus-Ammonellipsites* Zone. Subfissile, dark mudstones and laminated calcisiltites, which overlie the member, are exposed at a small tributary junction in the right bank. The Buckbanks Sandstone is exposed in this tributary and farther upstream in a meander of Leagram Brook near Buckbanks Barn [6354 4493], where the member is much thinner (0.8 m) than on the southern limb of the anticline. Upstream from the tributary intersection, dark subfissile mudstones and laminated calcisiltites predominate. A bullion band in the right bank [6352 4504] yields goniatite spat, together with a fine-grained packstone layer containing foraminifera including stunted *Glomodiscus* sp. and *Uralodiscus* sp., suggesting the Cf4β Subzone. Further upstream in the left bank [6348 4508], 30 m stratigraphically above the previous locality, a lenticular channel, about 1m thick and 2 m wide, is cut into mudstone and infilled with coarse-grained packstone. Foraminifera indicative of the Cf4γ Subzone are abundant and include *Endothyra* spp., *Eoparastaffella* sp., *Florenella* sp., forchiinids, *Glomodiscus miloni*, *Latiendothyranopsis menneri*, *Mediocris* sp., *Paraarchaediscus* stage *involutus*, *P. stilus*, *Planoarchaediscus* sp., *Plectogyranopsis ampla*, *Septabrunsiina* sp., *Tetrataxis* sp., *Valvulinella* sp. and *Viseidiscus* sp. (Plate 1h), together with algae including kamaeniids, *Koninckopora inflata*, *Salebra* sp. and stacheiinids. The Dunbarella Bed is exposed on the left bank [6333 4512]; it is about 5 cm thick and comprises fissile black mudstones with a band of small laminated bullions at the base. Some strike-slip movement has disturbed the bed. The bivalves *Dunbarella* and *Pteronites* are abundant. Some 6 m below the bed, blocky mudstones in the right bank yield small pyritised nomismoceratid ammonoids. Above the Dunbarella Bed, subfissile dark mudstones are gradually succeeded by bioturbated dark grey, blocky, calcareous mudstone and thin argillaceous wackestone. Towards the top of the formation [6323 4514] the beds become paler in colour, and the interbedded argillaceous wackestones become more numerous and transitional in character towards the overlying Hodderense Limestone Formation.

Exposures in the middle and upper parts of the formation occur in a stream [6317 4391 to 4382 6360] south-east of Throstle Nest; the sequence is similar to that in Leagram Brook, but is much less well exposed.

Plantation Farm Anticline

Good exposures, straddling the core of the anticline, occur in the gorge of the River Hodder [6631 4330 to 6712 4327] at Limekiln Wood, which is the type locality for the Limekiln Wood Limestone Member (Riley 1990a). The base of the member is not seen, but a 90 m-thick sequence is repeatedly exposed in isoclinal and box folds on both limbs of the anticline. Graded, fine- to very coarse-grained, crinoidal, cherty packstones predominate, interbedded with various amounts of micromicaceous, dark, blocky and subfissile mudstone. The packstones show erosive bases and internal, discontinuous, subparallel lamination; some beds are composite. Intraformational mudstone rip-up clasts and extraformational intraclasts of Waulsortian limestone commonly occur; the latter may be abundant and form breccias with interlocking stylolitic clast boundaries. Waulsortian clasts and crinoid debris also occur in some of the mudstone beds, and are taken to represent debris flows. Macrofaunas are abundant in the limestones, comprising hashes of fenestellid bryozoans and fragmentary brachiopods (chonetoids, orthids, orthotetoids, productoids and spiriferoids). The trilobite *Phillipsia gemmulifera*, reworked from the Clitheroe Limestone Formation, is also present. Local concentrations of silicified calyces of the crinoids *Actinocrinites* and cf. *Rhodocrinites* also occur. A mudstone debris flow from near the top of the member [6631 4330] has a particularly rich assemblage of Chadian crinoids including *Actinocrinites* sp. (Plate 1a–c), *Amphoracrinus* sp., *Bollandocrinus* sp., *Gilbertsocrinus* sp., *Platycrinites* sp., *Pimlicocrinus* sp. and *Pleurocrinus* sp., associated with the trilobite *Bollandia* cf. *persephone*. Foraminifera from the packstones include *Eblanaia michoti* and *Eotextularia diversa*, and the conodonts *Polygnathus bischoffi* and *Spathognathodus pulcher* are also present.

The faunas of the interbedded mudstones are much less diverse than, and different from those of the packstones and debris flows; they include nuculoid bivalves, indeterminate cephalopod debris and the trilobite *Latibole* sp. nov.. From Limekiln Wood, upstream to New Plantation [6567 4324], discontinuous exposures in the Phynis Mudstone Member occupy a complex syncline intersected by wrench faults and minor folds. Original bedding is largely obscured by cleavage, except at an exposure in the left bank of the river [6621 4332] where cleavage refraction is seen in diffusely bedded wackestones interbedded with blocky, finely micaceous mudstones. A thin development of the Limekiln Wood Limestone is seen in a steep plunging anticlinal fold in the left bank of the river at New Plantation [6564 4325]. Small overgrown outcrops adjacent to the path above the river at this point show Waulsortian limestone in the anticlinal core. It is not clear whether this limestone is in situ within or below the Hodder Mudstone Formation, or whether the outcrop represents part of a large detached block, or olistolith, within the Limekiln Wood Limestone. Upstream from New Plantation, towards Doeford Bridge [6507 4311], isolated outcrops of cleaved and folded Phynis Mudstone occur. Immediately upstream from the bridge, on the left bank, contorted and cleaved Phynis Mudstone is displaced by the Doeford Fault against folded uncleaved mudstones and thinly bedded, laminated calcisiltites in the middle and upper part of the formation. Discontinuous exposure is seen upstream from this point, in both banks, and comprises folded and sheared mudstones in the upper part of the formation. The junction with the overlying Hodderense Limestone Formation is seen in the left bank [6471 4344]. Some folding in this section appears to be due to syndepositional slumping.

Although minor folding and faulting occur, the southern limb of the Plantation Farm Anticline is uncleaved and less tectonised than the northern limb. Downstream from Limekiln Wood, the Limekiln Wood Limestone is overlain by the Phynis Mudstone Member, which is here estimated to be around 350 m thick. There are numerous outcrops in the river [6712 4325 to 6760 4297] north of Amridding Wood, where the member is succeeded by thin laminated calcisiltites and subfissile mudstones. The base of the Chaigley Limestone Member is exposed further downstream in the left bank [6789 4315] at Paper Mill Wood, which is the type section (Riley 1990b). Downstream from this point, successively thicker beds of graded, fine- to coarse-grained, cherty packstones become prevalent and exposure continues to the top of the member at Paper Mill Pool [6800 4258]. Erosive bases, mudstone rip-up clasts, reworked wackestone and phosphate nodules, and discontinuous subparallel lamination are common features. Some beds are composite, and slumping involving one or more beds also occurs. Interbedded mudstones are commonly fissile and may be silty, with septarian ironstone concretions and comminuted plant debris. The limestones are richly fossiliferous, with silicified hashes of fragmentary brachiopods, heterocorals, solitary rugose corals and rare caninioids, together with echinoid, crinoid and bryozoan debris. Ammonoids characteristic of the *Bollandites-Bollandoceras* Zone occur and include *Bollandites* sp., *Bollandoceras* sp., *Dzhaprakoceras* cf. *hispanicum*, *Hammatocyclus* sp. (Plate 1f,g), and *Merocanites* sp.. Foraminifera from the base of the member indicate a Cf4β–γ Sub-zone age and include *Eoparastaffella* sp., *Florenella* sp., *Glomodiscus miloni*, lituotubellids, *Mediocris breviscula.*, *Uralodiscus* sp. *Valvulinella* sp. and cf. *Viseidiscus* sp., together with algae including kamaeniids, *Koninckopora inflata* and *Salebra sibirica*. Exposures occur in nearby tributaries in Broad Meadow Wood [6803 4338], where the Dunbarella Bed is seen with its characteristic bivalve fauna (Plate 1r). Adjacent graded packstone beds contain a rich algal/foraminiferal Cf4β–γ Subzone assemblage, together with the conodont *Gnathodus pseudosemiglaber*. The upper part of the Chaigley Limestone Member (Plate 5) is well exposed in the River Hodder between Paper Mill Pool and Buck Hill, where the junction with overlying slumped, blocky mudstones is seen at the margin of the district [6840 4275]; exposure continues downstream into the adjacent Clitheroe district.

Another section through the core of the Plantation Farm Anticline occurs in Mill Brook [6732 4333 to 6691 4401], south-east of Lees House Farm. This exposes the Limekiln Wood Limestone Member and the overlying Phynis Mudstone Member in a secton parallel to that in the River Hodder at Limekiln Wood. The north-western limb of the anticline is cut by the Doeford Fault, which brings the upper part of the Chaigley Limestone Member against cleaved Phynis Mudstone Member [6690 4402].

Whitewell Anticline

On the northern limb, the oldest strata in the formation occur in a stream section [6485 4790 to 6472 4791] north-west of Lower Fence Wood, near Whitemore Knot. These comprise blocky, dark, calcareous, finely micaceous mudstones of the Phynis Mudstone, which are exposed up to a small waterfall, where a prominent bullion band occurs near the top of the member. The lip of the waterfall is composed of the basal beds of the overlying Whitemore

Limestone Member, for which this section is the type locality (Riley, 1990b). These beds consist of pale to dark grey, argillaceous, bioturbated wackestones and platy calcareous mudstones. They are richly fossiliferous and include a macrofauna of solitary corals, nuculoids, pectinoids, gastropods, orthocone and stroboceratid nautiloids, *Ammonellipsites* ex gr. *kochi*, *Dzhaprakoceras* sp., *Eonomismoceras* sp., *Merocanites* sp., *Weania gitarraeformis* and crinoid debris. The ammonoids are indicative of a high horizon within the *Fasciperycyclus-Ammonellipsites* Zone. Near the top of the section, graded beds of erosive-based, medium-grained packstone become common. These contain brachiopod hashes and fragmentary trilobites including *Cummingella* sp. nov., *Eocyphinium* sp., *Phillibole nitidus*, *Reediella stubblefieldi* and *Weania gitarraeformis*. Foraminifera indicative of the Cf4α2 Subzone include *Brunsia* sp., *Endothyra* sp., *Eoparastaffella* sp. (Plate 1o,q), *Eotextularia diversa* (Plate 1m) and *Florenella* sp., together with algae including kamaeniids and monolaminar *Koninckopora* sp., and the conodont *Gnathodus pseudosemiglaber*.

The contact with the overlying Leagram Mudstone Member is not exposed but, upstream from the boundary fence, there are isolated outcrops of mudstone lying in the middle part of the formation. South-westwards, in small outcrops [6474 4781; 6468 4767] near Whitemore Knot, a lateral facies change to crinoidal packstones and grainstones takes place, similar to that of the Limekiln Wood Limestone, but too limited in extent to be separately delineated on the map. The strike wraps around the Waulsortian limestone and similar outcrops are seen in an unnamed stream and adjacent hillside [6456 4746] immediately south of Whitemore Knot. Lithologies comprise coarsely crinoidal mudstones and packstones with an abundant, late Chadian macrofauna including *Cladochonus* sp., *Emmonsia parasitica*, *Rylstonia* sp., fenestellid bryozoans, orthotetoids, productoids, *Reticularia lineata*, *Rhipidomella mitchelli*, *Schizophoria connivens*, *S. resupinata*, *Spirifer* sp., *Anthraconeilo* sp, *Phestia* spp., *Sanguinolites* sp., *Dzhaprakoceras* sp., *Brachymetopus maccoyi*, *Cummingella* sp., *Latibole* sp. nov., *Phillibole nitidus*, *Amphoracrinus gigas*, *Cyathocrinites* sp., *Platycrinites* sp., *Pleurocrinus* sp. and *Mesoblastus* sp.. Immediately west of Whitemore Knot, in a swallow hole [6451 4765], the Limekiln Wood Limestone Member

Plate 5 Photomicrograph of a limestone turbidite from the Chaigley Limestone (Arundian), showing multistorey graded units.

The basal unit contains conches of the ammonoid *Hammatocyclus* reworked from an underlying bed of hemipelagic mudstone. Fine quartz sand is present near the top of the basal unit. The upper unit rests with sharp contact on the basal unit and is rich in crinoid, foraminiferal and algal debris derived from a shallow-water carbonate environment. Imbricate clasts of reworked hemipelagic mudstone also occur. Specimen no. RO 5085, from left bank of River Hodder [6830 4270], near Buck Hill.

is 4.8 m thick and consists of coarsely crinoidal packstones and thin mudstones, resting with sharp erosive contact on the Waulsortian Clitheroe Limestone. Above these basal beds, mudstones become more prominent and crinoids become sparse as the change into diffusely bedded, argillaceous wackestones and mudstones of the Leagram Mudstone Member occurs. A new, undescribed species of trilobite is present, together with *Winterbergia* cf. *hahnorum* and the conodonts *Gnathodus pseudosemiglaber* and *Polygnathus bischoffi*, confirming correlation with the Leagram Mudstone in the Clitheroe district. Near New Hey [6462 4861], a small outcrop of the Rain Gill Limestone Member is present, consisting of slumped, argillaceous packstone; otherwise exposure in the higher parts of the formation is poor.

On the southern limb of the Whitewell Anticline the Limekiln Wood Limestone Member contains spectacular boulder beds and breccias of derived Waulsortian limestone, which drape around and onlap against the Clitheroe Limestone Formation. Boulder beds on the valley side [6530 4447] south-west of Ing Wood form a strong feature and there are etched exposures in the River Hodder [6520 4452], nearby. A disused quarry to the north-east [6563 4486] exposes several faces showing interlocking boulders of Waulsortian limestone, some over 2 m across, embedded in crinoidal mudstone.

Thornley Anticline

On the south-eastern limb of the Thornley Anticline, the Limekiln Wood Limestone Member is well exposed at the disused Arbour Quarry [6200 4070], where the section extends for about 200 m along the south face, which is up to 5.0 m high. Composed entirely of subrounded boulders, this is one of the best examples of a boulder bed in the Craven Basin. The bed overlies Waulsortian limestones of the Clitheroe Limestone Formation, which outcrop in the central and northern parts of the quarry (p.19). The deposit was described by Parkinson (1935, pp.113–116), who noted boulders over 'six feet (1.83 m) long in diameter'. Boulders consist predominantly of unsorted clasts of Waulsortian limestone set in a mixed shaly mudstone and wackestone matrix that tends to weather back between the clasts, revealing their mainly subrounded shape (Plate 6).

Plate 6 Limestone boulder bed at Arbour Quarry [618 406] developed in the Limekiln Wood Limestone Member (late Chadian). Subrounded boulders of wackestone up to 0.5 m diameter were probably derived from a Waulsortian limestone exposed in the same quarry. The boulders are set in a mixed dark mudstone and rubbly limestone matrix (A14739).

The clasts appear to show a diverse orientation with respect to their original fabric. At one point in the face, there is a sharp, steep contact between a part of the boulder bed which is dominated by Waulsortian clasts and a part with clasts of vertically orientated shale and dark grey argillaceous wackestone of the Leagram Mudstone Member, yielding *Merocanites* sp. and *Latibole* sp. nov.. This suggests that the boulder bed is a product of more than one episode of erosion.

Parkinson (1935) reported that "a boulder bed of small extent" is present in the Leach House Quarry ("Bedlam Quarry") [628 412]. The exposure was not located during the present survey and is probably now overgrown, but 4.0 m of dark mudstone and argillaceous wackestone interbedded in equal proportion, possibly belonging to the Leagram Mudstone Member, are exposed at [6289 4117], where they overlie the Waulsortian limestone.

The Limekiln Wood Limestone Member is exposed in several isolated and overgrown faces [e.g. 6349 4278; 6353 4281] in a disused, partly flooded quarry at High Head Wood, north-west of Gibbon Bridge. The exposures, which are affected by folding and faulting, show sharp-based, graded beds of medium- to coarse-grained, crinoidal packstone interbedded with cleaved, crinoidal, finely micaceous mudstones and conglomerates composed of subrounded clasts of Waulsortian limestone, up to 0.25 m in diameter, set in a shaly matrix. The member is also exposed nearby to the east, in Leagram Brook [6374 4284 to 6377 4281], where there is a strongly folded sequence, about 10 m thick, of graded beds with crinoidal fine- to coarse-grained packstones interbedded with thin mudstones.

The Chaigley Limestone Member is exposed in small stream sections [6362 4142; 6365 4155; 6404 4202] between Thornley Hall and Carr Side Farm, represented by thin-bedded, flaggy packstone and mudstone. A small exposure of thin-bedded mudstone and dark, finely bioclastic packstones near the Waulsortian limestone at Lea House probably also represents this member, since it yields a diverse Cf4γ Subzone foraminiferal assemblage including *Glomodiscus* sp., *Paraarchaediscus* stage *involutus, Pseudolituotubella* sp. and *Uralodiscus* cf. *rotundus*, together with algae including kamaenids, *Koninckopora inflata* and stacheiinids.

Higher beds in the formation are limited to small outcrops of mudstone, with thin-bedded laminated calcisiltites, seen in stream sections [6322 4107] around Thornley Hall, and [6425 4194; 6417 4164] near Carr Side Farm.

On the north-western limb of the anticline, the lowest visible beds occur [6066 4114] near Crow Trees Farm, where 11.5 m of Buckbanks Sandstone Member are exposed (base not seen) in a small stream. The medium-grained, massive sandstone is grey to pale grey-brown, hard, and calcareous where fresh, but ochreous where weathered. A representative sample (E 61937), analysed by B Humphreys, has the composition of an orthoquartzite, with 84 per cent quartz including overgrowths, and a complete absence of feldspar (Appendix 2, Tables 3, 4 and 5). Chlorite, both authigenic and detrital, is a significant component of the matrix. The sandstone is overlain by 3.5 m of partly exposed grey, shaly, weathered mudstone. Foraminifera from thin limestones in these overlying beds belong to the Cf4γ Subzone and include *Eoparastaffella* sp., cf. *Glomodiscus* sp., cf. *Latiendothyranopsis* sp., and *Paraarchaediscus* stage *involutus*. One bed contained isolated radial-spar ooliths and peloids replaced by chert.

The Chaigley Limestone Member occurs in a small ditch exposure [6271 4240] 430 m north of Pale Farm; it consists of grey, medium-grained packstone with Cf4γ Subzone foraminifera including *Latiendothyranopsis* sp., *Paraarchaediscus* stage *involutus* and *Uralodiscus* sp.. The ditch traverses a north-east-trending feature which is thought to be produced by the limestone.

The westernmost exposure of the Hodder Mudstone Formation in the district is represented by the Chaigley Limestone; it occurs next to the Grimsargh Fault in the bed of Westfield Brook [5506

3796]. Here, 11.0 m of partly dolomitised fine- to medium-grained cherty packstones, with graded beds up to 0.4 m thick, are interbedded with calcareous siltstones and mudstones containing *Dzhaprakoceras* sp. and trilobite debris. Foraminifera obtained from the packstones, indicative of the Cf4g Subzone, include *Eoparastaffella* sp., *Eotextularia diversa, Glomodiscus oblongus, Paraarchaediscus* stage *involutus* and *Uralodiscus* sp.. A borehole (SD 53 NW/1) near Benson's House, 1.8 km to the south-west, provides some indication of the extent of the subdrift outcrop in that direction. Here some 52.4 m of thickly interbedded dark mudstone and white veined limestone were penetrated beneath 38.4 m of drift.

Exposures of beds high in the formation are limited to small streams around Crow Trees Farm [6041 4117] and near Little Mill Cottage [6107 4159].

Dunsop Bridge area

Exposures are poor in this area and are largly confined to isolated stream exposures south-east of Burholme Moor. One of these [6647 4792] exposes fissile mudstone and thin laminated calcisiltites with fragments of the bivalve *Dunbarella*, suggesting proximity to the Dunbarella Bed.

Slaidburn Anticline

On the north-western limb, the best sections are seen in the adjacent Lancaster and Settle districts. However, the base of the formation is exposed in Rough Syke [6716 5089], where 7.5 m of limestone breccias and packstones, representing the Limekiln Wood Limestone Member, rest with erosive contact upon the Clitheroe Limestone Formation (Thornton Limestone Member). The Rain Gill Limestone Member forms a strong bench feature, which is incised by a stream south-west of Oxenhurst, affording a good, but folded section [6715 5170] through the lower part of the member.

Exposure on the south-eastern limb of the anticline is restricted in this district to the River Hodder near Knowlmere. The Phynis Mudstone Member floors a large plunge pool [6840 4962] immediately downstream from a natural weir formed by Waulsortian limestones of the underlying Clitheroe Limestone Formation (p.15); further downstream, in the left bank, platy calcareous mudstones, interbedded with argillaceous wackestones and crinoidal packstones, are exposed. These beds belong to the Whitemore Limestone Member and yield the trilobite *Weania gitarraeformis*.

Sykes Anticline

The Hetton Beck Limestone Member forms the core of the anticline, which is exposed in Trough Brook, near the waterworks. The main exposure, about 85 m thick, is nearby on the north-western limb, in disused quarries to the east [6284 5188] and west [6266 5185] of the Trough of Bowland road, together with further exposures in adjacent small hillside scars. The member consists of argillaceous, fine- to coarse-grained packstones and thin, shaly mudstone interbeds. Macrofossils are abundant at some horizons and are rendered conspicuous by selective silicification and weathering, with numerous crinoids, bryozoa and brachiopods. Corals are prominent in the middle parts of the sequence, with large *Syringopora* colonies and solitary rugosans. Poorly bedded limestone units, up to 1.3 m thick, occur particularly in the middle part of the section. Some beds appear to contain intraformational lithoclasts, indicating reworking. Much of the internal bedding fabric of the limestone has been obliterated because of intense bioturbation; however, higher beds, particularly in the eastern quarry, are more thinly bedded and more coarsely crinoidal, and show sharp erosive bases and graded bedding. Soft-sediment slumping is common in the upper part of the sequence, especially in the eastern quarry; it has been described in detail by Gawthorpe and Clemmey (1985). Particularly conspicuous here is a large channel filled with slumped, graded packstone beds. Foraminifera and algae representative of the Cf4α2 Subzone are abundant and well preserved; they include the foraminifera *Brunsia spirillinoides*, *Dainella* sp., *Eoparastaffella* sp., *Eotextularia diversa*, *Florenella* sp., *Globoendothyra* sp., *Mediocris mediocris*, *Plectogyranopsis* sp., *Pseudotaxis micra*, *Septabrunsiina* sp., *Tetrataxis* sp., and *Tournayellina* aff. *beata*, and the algae *Girvanella* sp., kamaenids and both bilaminar and monolaminar *Koninckopora* spp.

In the highest parts of the quarry sections, cherts appear as discrete units. Relict sedimentary features, such as crinoid bioclasts, clearly demonstrate the replacement nature of the silicification, which locally extends throughout the formation into overlying strata and is associated with mineralisation (Chapter eight).

Above the Hetton Beck Limestone, the formation consists of dark subfissile calcareous mudstone with thin beds of calcisiltite showing subparallel lamination. These beds are best exposed in the adjacent Lancaster district but small exposures, including the overlying Hodderense Limestone Formation, occur in a stream near Sykes Farm [6313 5168].

Hodderense Limestone Formation (Holkerian)

The unit was first recognised and mapped by Parkinson (1926) in the Clitheroe Anticline to the south of the district, under the name Beyrichoceras hodderense Beds. Parkinson (1935, 1936) later mapped its outcrop in the Whitewell and Slaidburn areas. Earp et al. (1961), during the resurvey of the Clitheroe district, applied the term Bollandoceras hodderense Beds, thus accommodating Bisat's (1952) generic revision of *Beyrichoceras*. Fewtrell and Smith (1980) included these strata in their Worston Shale Formation, whereas Arthurton et al. (1988) included them in the Pendleside Limestone Formation. Riley (1990b) introduced the term Hodderense Limestone Formation and defined a stratotype in the River Hodder, near Stonyhurst, in the adjacent Clitheroe district. The formation lies within the Worston Shale Group, conformably between the Hodder Mudstone Formation below and the Pendleside Limestone Formation above.

Although relatively thin, usually between 5 and 15 m thick, it is the most widespread and uniform lithostratigraphical unit within the Worston Shale Group. The formation is present over the entire Dinantian outcrop in the present district and is probably widespread in the concealed areas also. The formation produces poor features and hence precise mapping relies on surface exposure and borehole sections. For this reason much of the outcrop as mapped is conjectural and is linked to the position of the overlying Pendleside Limestone Formation.

The ammonoid *Bollandoceras hodderense* is locally abundant, but it is the unique lithology which is the most conspicuous feature of the formation, which comprises thin beds of pale grey to cream weathering, porcellanous wackestone and floatstone, interbedded with thin beds of predominantly pale blue to olive-grey mudstones. Thicker beds of dark, blocky mudstone may occur, and in the Cow Ark Anticline thin beds of dark, argillaceous, medium-grained sandstone are also present. Both the mudstones and wackestones are, in places, chloritic. Irregularly shaped, dark bluish grey micritic nodules, less than 2 cm in diameter, are widespread, sometimes replacing cephalopods and other bioclasts. The nodules are structureless internally, even when viewed in polished section under a luminoscope. The nodules replace and protect ammonoids from compaction, suggesting that they formed at an early stage in diagenesis. This interpretation is further supported by signs of intraformational winnowing and reworking, since the wackestones commonly contain lag layers composed almost entirely of a jumbled array of nodules. Some of these lags show crude normal and inverted grading. Omission surfaces occur, portrayed by erosion of the upper surfaces of large ammonoids such as *Merocanites*. Bioturbation is predominantly of *Chondrites* type, although locally intense concentrations of *Thalassinoides* burrows are also present.

The benthonic fauna represents a community adapted to low oxygen levels (dysaerobic) and comprises solitary rugose corals, inadunate crinoid debris, microcrinoids, ophiuroid denticles, *Lingula* sp., *Orbiculoidea* sp. *Crurithyris* sp., *Nudirostra* spp., small productoids including *Linoproductus* and *Chonetipustula*, *Retispira*, nuculoid bivalves and *Latibole* sp. nov. (different from that occurring in the Hodder Mudstone Formation). Free-swimming and floating (nektopelagic) faunas are the most conspicuous, however, and include orthoconic and stroboceratid nautiloids, and ammonoids representative of the upper part of the *Bollandites-Bollandoceras* Zone (Riley, 1991), including *Bollandoceras hodderense*, *Bollandites* sp. nov., *Dimorphoceras* sp. nov. and *Merocanites* cf. *applanatus*. Conodonts, dominated by *Gnathodus girtyi*, are abundant in some beds. There is little variation in the relative proportions of faunal components between localities, although fossil abundance does vary.

Details

Sykes Anticline

The formation is poorly exposed in a stream near Sykes Farm [6313 5165], on the south-eastern limb of the anticline. Here, there are loose blocks of porcellanous creamy wackestones with characteristic dark bluish grey, micritic nodules, which are likely to have yielded the *Merocanites* referred to by Moseley (1954). The formation is well exposed on the north-western limb of the anticline in the adjacent Lancaster district.

Plantation Farm Anticline

On the north-western limb of this anticline, good exposures occur on the left bank of the River Hodder upstream from Doeford Bridge [6470 4345], where the formation is 7 m thick. The contact with the contiguous Hodder Mudstone and Pendleside Limestone formations is also seen. Mill Brook exposes two sections through the formation [6755 4482; 6706 4437] near Micklehurst Farm. In the northernmost exposure the section is complicated by minor folding, and the junction with contiguous formations is also exposed. Further partial exposure occurs in a stream [6624 4445] south-west of Higher Lees Farm.

Whitewell Anticline

The best exposure in the Whitewell Anticline occurs on the south-eastern limb, in a stream section at Pale Wood [6579 4435], west-south-west of Higher Lees Farm; contiguous formations are also seen.

On the north-western limb of the anticline, the uppermost beds and the overlying Pendleside Limestone Formation are seen in a small stream exposure [6493 4903] near New Hey Farm. Some of the wackestone beds here are so rich in lags of bluish grey micritic nodules that they form breccias.

Cow Ark Anticline

The formation is not exposed in the Cow Ark Anticline and is known only from boreholes, including SD 64 NE/5,7,16,19,20 and 23 (Figure 6 and 7). Some of the wackestones contain thick lags of bluish grey micritic nodules (Plate 7a) and a thin, dark, silty, poorly sorted, medium-grained sandstone was present in SD 64 NE/23.

Throstle Nest Anticline

The best exposure of the formation in the district occurs on the north-western limb of this anticline, in the deep ravine of Leagram Brook [6322 4513], south-east of Park Style, where the relatively shallow dips have provided a long section; contiguous formations are also seen. The formation is relatively shaly and richly fossiliferous. Upper beds in the formation are also partly exposed in Hell Clough [6360 4631], north of Lickhurst Farm.

The only exposure on the south-eastern limb occurs in a stream [6360 4382] south-east of Dairy Barn; only the lowest beds are seen, overlying the Hodder Mudstone Formation, before exposure is lost beneath the drift.

Thornley Anticline

Small exposures are present on the southern limb of the Thornley Anticline in a ditch [6259 4046] south-west of Arbour Farm and in stream south of Carr Side Farm [6422 4153]. The formation is not exposed on the north-western limb.

Pendleside Limestone Formation (Holkerian – late Asbian)

The Pendleside Limestone was first recognised by Tiddeman (1889) and the name was subsequently used by Parkinson (1926, 1935, 1936) and Earp et al. (1961). It is broadly equivalent to the "Pendleside Shales with Limestone" of Hind and Howe (1901). Formational status was first given by Fewtrell and Smith (1980), who defined a stratotype at Pendle Hill in the adjacent Clitheroe district. In the Settle district, Arthurton et al. (1988), for practical mapping purposes, took the base much lower than previous authors by including the Bollandoceras hodderense Beds (Hodderense Limestone Formation) and beds in the upper part of the Hodder Mudstone Formation. Gawthorpe (1987) and Arthurton et al. (1988) also included limestones which Riley (1990b), considered to belong to the overlying Bowland Shale Group. During the present survey these higher limestones, including the Ravensholme and Park Style Limestones, have been treated as members within the Bowland Shale. The Pendleside Limestone Formation in the present district is thus bounded conformably by the Hodderense Limestone Formation below and by the Bowland Shale Group above. Metcalfe (1981) applied the term Rad Brook Beds to shaly strata which occur locally, particularly in the lower part of the formation, and which Riley (1990b) later renamed the Rad Brook Mudstone Member after the stratotype at Rad Brook, on Pendle Hill.

The formation is widespread and gives rise to distinctive topographical features. Thickness is highly variable, ranging in this district from 10 to 50 m. Limestone turbidites dominate, comprising beds of detrital, fine- to coarse-grained packstone. Bedding is mainly planar, with both thin- and thick-bedded units commonly present as composite beds. Graded beds with sharp erosive bases, tool and prod marks, small- and large-scale soft sediment deformation, parallel and discontinuous subparallel lamination, and intra-formational mudstone conglomerates are common. Limestone breccias occur locally, as in the Sykes Anticline; they comprise clasts of limestone derived from the shallow-water environments which surrounded the Craven Basin. Other derived extrabasinal components, often visible in petrographic thin-sections, include radial-spar ooliths and coated grains; these may be solitary or clumped as aggregate grains, embedded in a bioclastic packstone matrix. Bioclasts are mainly allochthonous and comprise disarticulated and fragmentary brachiopod valves, crinoid ossicles, foraminifera and algal chips. In the lower part of the sequence, cream-coloured wackestones may occur; they are transitional in character to the underlying Hodderense Limestone Formation, but lack the characteristic grey nodules. Terrigenous silt is a minor component, and sandstone has been found only in boreholes in the eastern part of the Cow Ark Anticline. Mudstones are predominantly blocky or platy, calcareous, pale to dark grey and laminated.

The Rad Brook Mudstone Member locally replaces the limestones, especially at the base of the formation, and may account for the whole thickness of the formation. The member is up to 48 m thick and comprises mudstones with only minor interbeds of laminated calcisiltite, fine-grained packstone and wackestone. Intervals of shaly strata may also separate limestone-dominated parts of the formation into leaves. Bioturbation is a conspicuous feature in both the mudstones and limestones, and may disrupt the internal bedding of the latter. Dolomitisation is common, particularly in the packstones. Silicification is widespread and may be selective, affecting only bioclasts, or it may form lenticles and bands of chert within limestone beds. In the Sykes Anticline and the northern limb of the Throstle Nest Anticline, the entire formation is locally replaced by chert. The mudstones contain faunal groups similar to those present in the Hodder Mudstone and Hodderense Limestone formations, which indicate a continuation of dysaerobic conditions. They include solitary rugose corals, the brachiopods *Chonetipustula*, *Rugosochonetes* and *Tornquistia*, nuculoid and pectinoid bivalves, nautiloids, ammonoids and trilobites. An horizon with the dendroid graptolite *Callograptus* is present near the base of the formation (Hind, 1907). Trace fossils are particularly abundant and include *Chondrites*, *Cruziana*, *Helminthoidea*, *Helminthopsis*, *Planolites*, cf. *Thalassinoides*, *Zoophycos* and a vertical *Monocraterion*-like trace.

Details

Sykes Anticline

The only exposure of the formation in the Sykes Anticline within the district is in Penny Brook [6307 5151], near Sykes Farm, where 6 m of intensely bioturbated, pale cream-coloured wackestones overlie the Hodderense Limestone Formation. Upstream, the Pendleside Limestone is thrown down to the south-west by a fault.

0

-5

-10

-15

-20 cm

a)

b)

Plate 7 Slabbed core of a) graded nodule-bearing floatstone from the Hodderense Limestone Formation (Holkerian), showing a micrite matrix enclosing micrite nodules, some replacing ammonoid conches (*Bollandoceras hodderense*, arrowed). Borehole SD 64 NE/23. 48/24 to 48/71 m depth.
b) debris bed from the Pendleside Limestone Formation (Asbian), showing a breccia comprising reworked intraformational clasts of wackestone and fine packstone in a crinoidal matrix. Borehole SD 64 NE/20, 81.26 to 81.66 m depth.

Plantation Farm Anticline

On the north-western limb of this anticline, the lowest 24 m of the formation are exposed on the left bank of the River Hodder upsteam from Doeford Bridge [6468 4349]. The succession consists predominantly of cherty, fine-grained packstones and wackestones, but interbedded dark grey mudstones form about half the sequence in the upper 17 m of the section. Extensive strike sections in folded, fine- to coarse-grained, cherty packstones and mudstones occur in Mill Brook [6705 4441], west of Micklehurst Farm. Farther upstream [6750 4483 to 6747 4482], 450 m north-west of Micklehurst Farm, the brook has cut a ravine through the entire formation, providing one of the best sections in the district. A thin lithoclastic debris bed 8.25 m above the base of the formation yielded a large fragment of a palaeotextulariid foraminiferan referable to cf. *Climacammina*, suggesting a possible Cf6γ Subzone correlation within the *Neoarchaediscus* Zone (Figure 4). The presence of this subzone is confirmed by the occurrence of *Koskinobigenerina* in a 1.8 m-thick, medium- to coarse-grained cherty packstone, the topmost bed in the formation, which also yielded solitary and aggregrate radial-spar ooliths.

On the southern limb of the anticline, the formation forms a prominent bench feature along the foot of the Longridge Fell escarpment. Several sections occur in small streams [6835 4186; 6811 4187; 6814 4196] east of Walker Fold, as well as in a deeply excavated, disused quarry [6597 4175] near Head o'th Moor, where an extensive 20 m section of cherty packstones occurs.

Whitewell Anticline

The formation in the Whitewell Anticline is 30 to 35 m thick. Exposure is seen on the north-western limb in a gully [6447 4845] draining the backscar of a large landslip near New Hey Farm. The upper 4 m of the section, comprising bioturbated pale to dark grey mudstone, is underlain by 6 m of interbedded mudstones and packstones. A further 20 m of the formation is discontinuously exposed further down section, but is largely obscured by loose blocks. About 8 m of the lower part of the formation, comprising packstones, wackestones and mudstones, is well exposed in a fresh landslip scar [6443 4855] 500 m south-west of New Hey Farm. There is a highly disturbed exposure between the two branches of the Mellor Knoll Fault, east of New Hey Farm [6490 4906], where approximately 10 m of interbedded packstones and mudstones are seen.

Cow Ark Anticline

The formation is not exposed in this anticline and is known only from boreholes, including SD 64 NE/5,6,7,16,19,20 and 23 (Figures 6 and 7; Plate 7b). The proportion of limestone to mudstone varies considerably between boreholes, as does the formational thickness. In borehole SD 64 NE/7 the formation is about 15 m thick and is dominated by bioturbated mudstones, which are assigned to the Rad Brook Mudstone Member; there is a 2 m-thick intercalation of wackestones and a thin packstone in the middle part. In boreholes SD 64 NE/16 and 19, by contrast, the formation is about 37 m thick, with packstones and thinly bedded wackestones being prevalent in the lower and middle part, and mudstones being dominant in the upper part. Thin, medium-grained, argillaceous sandstones are unique to this area and are present in boreholes SD 64 NE/5,6,7,19 and 23.

Throstle Nest Anticline

The best exposure of the formation in the district occurs on the north-western limb of this anticline in Leagram Brook [6320 4515 to 6298 4511], south of Park Style, where gently folded strata are exposed in waterfalls and in a ravine section. The formation is 36 m thick and is represented by the Rad Brook Mudstone Member. Thin-bedded wackestones occur in the basal 2 m of the sequence and in a 2 m-thick unit 6 m below the top, which forms a small waterfall. The rest of the section is predominantly of cherty mudstone. A few thin (maximum 5 cm) layers of greenish grey clay of unknown origin occur towards the middle of the sequence. Macrofossils are rare, but crinoid debris, *Chonetipustula, Tornquistia* and crushed *Merocanites* were noted. At Hell Clough [6351 4633] the Rad Brook Mudstone Member appears to occupy the entire formation, but exposure is poor. The outcrop of the formation on the south-eastern limb of the anticline is conjectural since it is not exposed.

Chipping Anticline

The formation comprises the core of the Chipping Anticline. The best exposure is in Chipping Brook, at Chipping [6234 4333], where 12.4 m of well-bedded grey to dark grey packstones and wackestones are seen. The packstones are sharp-based, up to 0.59 m thick (average about 0.25 m) and show mainly parallel and some undulating lamination. Chert lenses are common and mudstones are present mainly as thin partings. Coated grains and isolated radial-spar ooliths were noted in thin section. Two samples yielded foraminifera, which suggest a Cf6α Subzone correlation, including bilaminar palaeotextulariids, *Mediocris* sp., *Paraarchaediscus* stage *angulatus, P. stilus, Plectogyranopsis convexa* and *Septabrunsiina* sp.. Associated algae include *Coelosporella* sp., *Koninckopora inflata, K. minuta* and stacheiinids.

There are further exposures of similar folded beds farther upstream just below a weir [6226 4342], and in a stream section [6175 4266] north of Radcliffe Hall.

Thornley Anticline

On the north-western limb, streams south-west of Chipping, at Wallclough [6057 4156], Astley House [6101 4190] and Radcliffe Hall 6183 4255], reveal the presence of limestone beneath the drift, but do not provide extensive sections. At Wallclough about 40 m of thinly bedded, dark grey packstones with common chert and mudstone intercalations are discontinuously exposed.

Farther south-west the formation is estimated to be at least 60 m thick. A stream section at Loudscales [5906 4026 to 5901 4030] shows a 56.38 m-thick sequence, about 20 per cent of which is obscured by alluvium. The upper 18 m shows interbedded dark grey calcareous mudstone and limestone, the former predominating. The limestones include packstones and wackestones, some in sharp-based graded beds up to 1.05 m thick. One such bed contains ooliths and bioclasts in the basal 3 cm and a large, allochthonous, cerioid *Lithostrotion* colony was noted in one of the packstones. Some slumped bedding and a few chert lenses are present. Limestone also predominates in the lower part of the sequence, again with some slumping and chert lenses. Bioturbation is conspicuous, especially in the mudstones. The mudstones yielded crushed *Merocanites* sp. with the brachiopod *Leptaenisca culmica* attached. Foraminifera, suggestive of the Cf6α Subzone, obtained from the packstones include *Endothyra* spp., *Paraarchaediscus* stage *angulatus, Planoarchaediscus* sp. and *Pseudommodiscus* sp.; they are associated with the alga *Koninckopora inflata*.

On the south-eastern limb the best sections occur in Cutlers Quarry [6239 4007] and the adjacent Dale House Quarry [6253 4022]. The combined succession, totalling 30.77 m, probably starts about 10 m above the base of the formation. Many characteristic features of the formation are displayed, including thick massive beds of dark grey wackestone and thinner graded wackestone/packstone beds with sharp bases and directional sole structures, slump bedding, common black chert lenses and nodules, and bioturbation. To the north-east, a stream [6340 4091] south of Thornley Hall provides a discontinuous section through the limestone and a more continuous one through the upper part of the Rad Brook Mudstone. A small disused quarry, just beyond the top of this section, partly exposes a 6 m-thick sequence of dark grey laminated and lenticular bedded limestone with internal slump folds at the top. A somewhat larger disused quarry east of Thornley Hall, near Brook House Farm [6395 4120], shows only poor exposure. However, an adjacent stream running along the south-west side of the quarry displays a 11.87 m-thick sequence of folded, dark grey packstones and wackestones interbedded with mudstones. Some of the packstones, which are up to 1.2 m thick, are laminated. The interbedded, dark, shaly mudstones probably make up nearly half the total sequence, but are poorly exposed. The section continues upstream into the overlying Bowland Shale Group. Farther southwest the formation can be traced along a bench feature to gully sections above Wheatley Farm [6211 3981 and 6204 3975], where scattered exposures of thinly bedded, fine-grained packstone with thin tabular cherts and subordinate mudstones occur. The formation is estimated to be around 60 m thick in this area.

The Pendleside Limestone was proved in borehole SD 53 SW/7 between 184.4 and 193.7 m depth, beneath Permo-Triassic cover. The reddened mudstones and calcisiltites, interbedded with limestone, contained the ammonoid *Merocanites* sp. associated with the trilobite *Latibole* sp. nov..

Slaidburn Anticline

In the district the formation is not exposed in the Slaidburn Anticline, although good sections are seen to the east of the district. On the north-western limb the formation is thought to be around 20 m thick and forms a poor bench feature along the valley side adjacent to Beatrix Fell.

BOWLAND SHALE GROUP (LATE ASBIAN–PENDLEIAN)

The Bowland Shale Group (Earp et al., 1961) comprises the Lower Bowland Shale and Upper Bowland Shale formations. It corresponds closely to the Bolland Shale of Phillips (1836), the Bowland Shales of Tiddeman (in Hull et al., 1875), the Bowland Shale Series of Parkinson (1926) and the Bowland Shale Formation of Fewtrell and Smith (1980). Since the Upper Bowland Shale Formation is of Namurian age it is dealt with in the succeeding chapter.

Lower Bowland Shale Formation (late Asbian–Brigantian)

The stratotype is at Little Mearley Clough on Pendle Hill in the Clitheroe district, which follows Fewtrell and Smith's (1980) proposal for the base of their Bowland Shale Formation; a description was given by Earp et al. (1961, pp.89–91). In the present district, as at the stratotype, the formation is conformable between the Pendleside Limestone Formation below and the Upper Bowland Shale Formation above. It varies in thickness from 55 to 400 m and consists predominantly of mudstone, with variable amounts of sandstone, limestone, limestone debris beds and breccias. The mudstones represent the background sediment and are the primary lithology by which the formation is recognised. They are black, calcareous, foetid and petroliferous. Pyrite is common along joints and sometimes replaces bioclasts. In-

ternal lamination is prevalent and the mudstones may be blocky or shaly. Fissile paper shales, which are so characteristic of the succeeding formation, are unusual. Interbedded limestones include argillaceous wackestones, packstones and breccias; sharp-based graded beds are common at some levels in the sequence. Nodular wackestones or "bullions", which result from localised early cementation of the mudstone, occur at discrete horizons. Fossils recovered from these nodules are undistorted by compaction and any fractures or cavities commonly bleed with mineral oil when freshly broken. The black mudstones at the base of the formation show a marked colour change from the pale and dark grey, locally olive and blue-grey colour of the Worston Shale Group, accompanied by a conspicuous reduction in bioturbation.

Thin limestones are scattered throughout the formation, and there is a major limestone, termed the Park Style Limestone Member, within the lower part. Its lower division is approximately coeval with the Ravensholme Limestone Member of the Clitheroe district (Earp et al., 1961) and it forms good features in the Chipping area, where it is best exposed. The type section is here taken in Leagram Brook (Figure 8), near Park Style [6294 4514 to 6280 4526]. The member thins away from the type section, where it is 84 m thick, to only 6.3 m at Hell Clough [6353 4640], north of Lickhurst Farm, 1.3 km to the north-east. Packstones and wackestones, which predominate, are interbedded with variable amounts of mudstone of typical Lower Bowland Shale Formation lithology. In the present district, the Ravensholme Limestone Member occurs only as a small outcrop on the south-east limb of the Sykes Anticline (Hughes, 1986).

Lower and upper subdivisions of the Park Style Limestone are recognisable in Leagram Brook. The former comprises the basal 17 m, in which the proportion of limestone to mudstone is roughly equal. Individual limestone beds range from from 8 to 70 cm thick (average 30 cm), are dark bluish grey and typically have sharp bases. Flute and groove casts, and sharp, undulatory tops also occur. These beds consist of medium- to coarse-grained packstone fining upwards into calcisiltite and eventually into argillaceous wackestone, which usually forms the bulk of each bed. Undulatory parallel lamination occurs in most beds, with less common convolute and slumped lamination. The interbedded mudstones are cherty, calcareous and platy or fissile. Some composite and subcentimetre-scale layers of upward-fining, very fine sand and silt are common at some levels. Thin layers of bioclastic debris also occur in places, consisting mainly of crinoid, brachiopod and coral fragments. Some of the packstones contain isolated and aggregrate grains of radial-spar ooliths. Bioturbation is absent.

The upper subdivision comprises the top 67 m of the member. The proportion of limestone to mudstone varies from 70 to 90 per cent. The limestones are wackestones consisting of grey, slightly argillaceous, siliceous micrite, locally weathering to a pale brownish grey. Beds range from 8 to 80 cm thick (average 35 cm), with generally sharp, slightly undulating bases and tops. Faint parallel lamination is visible in some beds. A 1.06 m bed of fine- to medium-grained packstone, with local undulatory lamination, occurs approximately 45 m below the top of the sequence. The interbedded

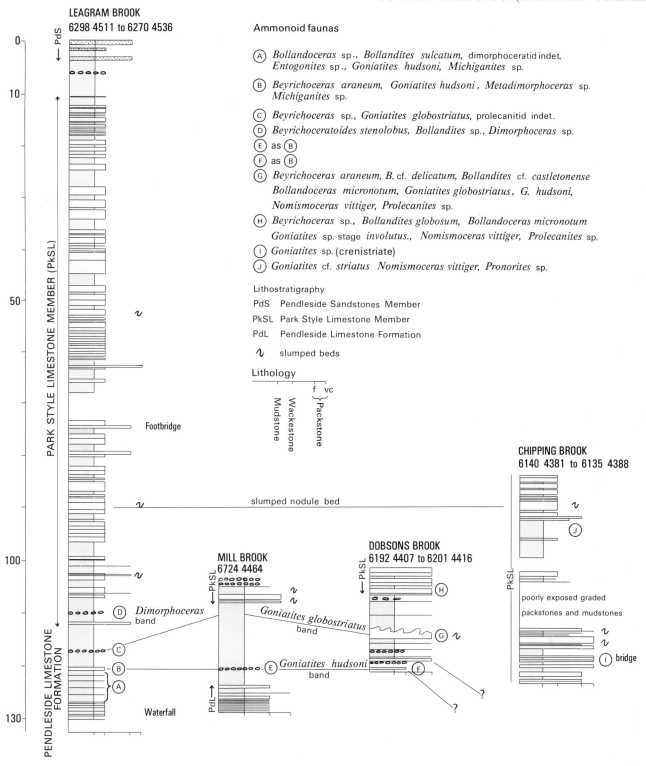

Ammonoid faunas

(A) *Bollandoceras* sp., *Bollandites sulcatum*, dimorphoceratid indet, *Entogonites* sp., *Goniatites hudsoni*, *Michiganites* sp.

(B) *Beyrichoceras araneum*, *Goniatites hudsoni*, *Metadimorphoceras* sp. *Michiganites* sp.

(C) *Beyrichoceras* sp., *Goniatites globostriatus*, prolecanitid indet.

(D) *Beyrichoceratoides stenolobus*, *Bollandites* sp., *Dimorphoceras* sp.

(E) as (B)

(F) as (B)

(G) *Beyrichoceras araneum*, *B.* cf. *delicatum*, *Bollandites* cf. *castletonense* *Bollandoceras micronotum*, *Goniatites globostriatus*, *G. hudsoni*, *Nomismoceras vittiger*, *Prolecanites* sp.

(H) *Beyrichoceras* sp., *Bollandites globosum*, *Bollandoceras micronotum* *Goniatites* sp. stage *involutus.*, *Nomismoceras vittiger*, *Prolecanites* sp.

(I) *Goniatites* sp. (crenistriate)

(J) *Goniatites* cf. *striatus* *Nomismoceras vittiger*, *Pronorites* sp.

Lithostratigraphy

PdS   Pendleside Sandstones Member
PkSL Park Style Limestone Member
PdL   Pendleside Limestone Formation

ᴨ slumped beds

Lithology

Figure 8   Sections in the lower part of the Lower Bowland Shale Formation.

dark mudstone is calcareous, blocky and strongly bioturbated at some levels, but individual trace fossils are blurred and diffuse. The packstones contain mainly crinoid, brachiopod and coral fragments, but the rest of this upper subdivision appears devoid of macrofossils.

Thin beds of sandstone occur at several levels in the formation, and there are also discrete levels where sandstone is dominant; these are recognised collectively as the Pendleside Sandstones Member. The member is equivalent to the "Pendleside Grit" shown on the one-inch-to-one-mile Old Series geological map of the Burnley Coalfield and adjacent areas (Sheet 92 SW). The unit was later renamed "Lower Yoredale Grit" in the accompanying memoir (Hull et al., 1875). Pendleside Grit was the name used by Parkinson (1926), which he later termed "Upper" and "Lower" Pendleside Sandstone (Parkinson, 1936). Pendleside Sandstone (Earp et al., 1961) and Pendleside Sandstones (Arthurton et al., 1988) are other names. The sections exposed at Red Syke [8103 4300] and Pendle Hill Brook [8095 4230], on the flank of Pendle Hill, as described by Earp et al. (1961, p.92), are here designated the stratotype.

Dominant lithologies are pale to dark grey sandstone (ochreous when weathered) comprising subangular quartz grains with minor amounts of plagioclase, alkali feldspar and lithic fragments, mostly polycrystalline quartz, interbedded with variable amounts of mudstone and siltstone. A few beds of sandstone are slightly calcareous and contain small fragmentary bioclasts, mainly crinoid ossicles. Noncalcareous sandstones contain no bioclastic material, but some have small yellowish white flecks, possibly composed of authigenic kaolinite, which are pseudomorphs after bioclastic material. Petrographical data from three samples of sandstone (E61938-61940), analysed by B Humphreys, are given in Appendix 2 (Tables 3,4 and 5). The sandstones show a variety of bedforms, which range from thin planarbedded to thick irregular-bedded units, and include argillaceous shaly intervals.

In the planar-bedded sandstones, grain size varies from fine to coarse grained and sorting is moderate to poor; there is commonly a fairly large admixture of silt and clay. Carbonaceous fragments and mudstone rip-up clasts are common, the latter typically being distributed along bedding planes. Bedding surfaces are near parallel and individual beds are 0.1–0.9 m (average 0.4 m) thick. Mudstone partings separate some beds, but the proportion of mudstone is less than 15 per cent. Internally, beds usually show slight normal grading, with convolute, planar-parallel and currentripple lamination developed locally towards the top. Sole structures occur on sandstone beds immediately overlying mudstone partings.

The irregularly bedded sandstones vary from medium to coarse grained and are poorly graded, individual beds being even grained. Carbonaceous fragments and mudstone ripup clasts are locally common. Bedding is moderately thick (0.2–1 m, average 0.5 m) and undulatory, with individual bedding surfaces commonly disappearing or amalgamating laterally over distances of a few metres. Most beds are structureless internally, but some display normal vertical grading. Impersistant mudstone partings occur, but are not common.

Shaly sequences of interbedded sandstone and mudstone have sandstone beds which resemble the planar-bedded sandstone type in terms of internal structure, but which are much thinner (0.03–0.6 m, average 0.25 m) and finer grained (very fine to medium). Sole structures, including groove, flute and load casts, are fairly common. The interbedded mudstone dominated intervals comprise between 15 and 70 per cent of the total thickness (average 50 per cent). The mudstone may be either calcareous and fissile, with thin layers of very fine sand or silt, or a very poorly sorted mixture of mud, silt and very fine to coarse sand, with abundant mica and carbonaceous fragments. A few beds of argillaceous wackestone occur within calcareous mudstone intercalations.

The Lower Bowland Shale Formation is generally very fossiliferous. Autochthonous benthonic (bottom-dwelling) faunas dominate the mudstones. They are adapted to dysaerobic (oxygen-poor) conditions and are less diverse than in the underlying Worston Shale Group which, together with the generally poor trace fossil content and petroliferous nature of the mudstones, suggests that oxygen depletion was more severe. Benthonic fossils are often preserved on single, or closely spaced bedding planes and are dominated by thin-shelled epibyssate bivalves, principally *Actinopteria persulcata*, *Caneyella* spp., *Dunbarella persimilis* and *Posidonia* spp.. Brachiopods are rare, but include smooth spiriferoids and *Nudirostra* sp.. Molluscs which lived in the substrate, such as nuculoid bivalves and bellerophontid gastropods, are also rare. Trilobites are common only in thin bands near the base of the formation. Nektopelagic (swimming and floating) faunas are dominated by both nautiloid and ammonoid cephalopods, conodonts and fish. Like the bivalves, these faunas tend to be concentrated in bands, usually of platy, calcareous mudstone similar to the "marine bands" in the overlying Namurian, but unlike them in that ammonoids also occur in the intervening mudstones. Packstones are composed predominantly of transported and fragmented shelly faunas derived from the shallow-water carbonate environments that surrounded the Craven Basin. They consist mainly of foraminifera, coral, crinoid, bryozoan and brachiopod debris. Some of the foraminifera and algae have been reworked from mid-Dinantian sediments.

Details

Sykes Anticline

The formation is best-exposed in the northern part of the anticline, which lies in the adjacent Lancaster district. The Pendleside Sandstones Member is poorly developed, but a 27 m-thick section, including beds low in the member and the underlying mudstones, is exposed in Swine Clough [6240 5190], north of Stake End. Several small gullies on the north-western face of Staple Oak Fell [around 6355 5160], 560 m east-north-east of Sykes, expose blocky and fissile mudstones in the P2c Ammonoid Subzone, associated with *Lyrogoniatites georgiensis*. The junction with the Upper Bowland Shale Formation is also seen.

Plantation Farm Anticline

On the north-western limb of this anticline a tightly folded sequence of Lower Bowland Shale, including the junction with the underlying Pendleside Limestone Formation, is well exposed in a strikecontrolled ravine cut by Mill Brook [6724 4463 to 6747 4485], north-west of Micklehurst. At least 35 m of the formation are represented, the top 4.6 m forming in the Park Style Limestone Member; the beds are exposed in a tight syncline between two tufa mounds and in a tributary waterfall entering Mill Brook from the north-west [6745 4487]. The entire section of Lower Bowland Shale lies within the late Asbian, B2a and B2b ammonoid subzones (Riley, 1990a) (Figure 4). The westernmost part of the section is the best exposed [6724 4463] and a graphic log is given in Figure 9. The Goniatites hudsoni and Goniatites globostriatus bands occur 5.05 m and 15.7 m above the base of the formation respectively. A cherty, medium-grained packstone, 0.6 m above the base, yielded *Pojarkovella* sp., indicating reworking of foraminifera from Holkerian or early Asbian strata. The basal beds of the Park Style Limestone comprise coarse-grained, slumped, foetid packstones. Solitary and aggregate grains of radial-spar ooliths were noted in thin section.

On the southern limb of the anticline, exposure is limited to the Pendleside Sandstones Member, which forms a prominent bench along the lower slope of the Longridge Fell escarpment. The member is about 100 m thick, but thins markedly to the south-east. It is seen in a small stream section near Walker Fold [6720 4161], where 1 m of brown-weathering, fine-grained, quartzose sandstone is exposed.

Whitewell Anticline

On the north-western limb the formation is about 85 m thick, but is generally poorly exposed. Thinning of both the Park Style and Pendleside Sandstones members is the main reason for the attenuation of the sequence. The Goniatites globostriatus band is exposed in the banks of an unnamed stream 250 m north-west of New Hey Farm [6446 4917] and lies some 8 m above the Pendleside Limestone Formation. Intervening strata consist of black mudstones. Partial exposure continues for 70 m upstream of the G. globostriatus band, where a few thin (less than 5 cm thick) packstone beds are interbedded with black mudstones. These may be equivalent to the thicker packstones which occur in the lower part of the Park Style Limestone Member to the south and south-west. Good exposures occur along the left bank of Langden Brook [6548 5002 to 6510 5017], where blocky, black mudstones and wackestone "bullions" are seen. Detailed fossil collecting was not undertaken, but the presence of large numbers of *Posidonia becheri* suggests a P1b or P1c Ammonoid Subzone correlation.

Strata in the middle part of the formation are exposed in a shallow gully, 750 m north-west of New Hey Farm [6420 4961], where black mudstones have yielded *Sudeticeras* sp., indicating an age no older than the P1c Ammonoid Subzone. These mudstones lie approximately 20 m below the base of the "middle leaf" of the Pendleside Sandstones. Only this middle leaf is represented in the area around New Hey Farm and Mellor Knoll; it is about 55 m thick, but to the north-east thins rapidly to only 5 m in a small tributary of Green Clough [6402 4971].

Exposures above the Pendleside Sandstones include a small section between the two branches of the Mellor Knoll Fault, 350 m north of New Hey Farm [6470 4934], where black mudstones yield *Posidonia trapezoedra* and indeterminate striatoid ammonoids, indicating an age range within the P1d to P2c ammonoid subzones. An horizon of wackestone bullions and black mudstone is exposed in the bank of Langden Brook [6505 5011], and shaly, black mudstones occur in a gully [6515 5004] on the hillside above. The presence of *Neoglyphioceras* sp. at the latter locality suggests a P2a–b Ammonoid Subzone correlation. A small exposure of black mudstone [6414 4917] in a gully 520 m west of New Hey Farm lies approximately 17 m above the top of the Pendleside Sandstones; the occurrence of *Sudeticeras splendens* indicates a P2b Ammonoid Subzone horizon.

About 600 m further north, at the confluence of Green Clough and Cherry Gutter [6393 4954], on the north-west face of Totridge

a 3.5 m-thick section of black mudstone yields *Sudeticeras* cf. *ordinatum*, suggesting an horizon within the upper part of the P2b Ammonoid Subzone. In a gully [6384 4941] 140 m upstream from the confluence with Cherry Gutter, a 3 m black mudstone band yields *Lyrogoniatites georgiensis*, indicative of the P2c Ammonoid Subzone. This represents the highest ammonoid horizon in the Lower Bowland Shale; it is overlain by 2 m of barren, black mudstone. The base of the overlying Upper Bowland Shale Formation is not exposed, but it probably occurs within a few metres of the top of the section.

Cow Ark Anticline

The formation is not exposed in the Cow Ark Anticline and is known only from boreholes including, SD 64 NE/6,7,16,17,19,20 and 23 (Figures 6 and 7). Because the sections are faulted, the thickness of the formation is not known; however, it is likely to be relatively thin, in the order of 50 m. Unfaulted contacts with the underlying Pendleside Limestone Formation occur in boreholes SD 64 NE/7, 16 and 19. In each case the junction is preserved in mudstone and associated with a loss of bioturbation and a colour change from pale or dark blue-olive-grey below the boundary to black above. Thin sandstones are present in the basal beds. These persist from earlier formations and are not included in the Pendlesides Sandstones Member, although a sandstone bed in borehole SD 64 NE/16, between 9.83 and 14.41 m depth, may belong to the member. Several ammonoid horizons (Figure 4) were recognised within the following subzones; B2a (SD 64 NE/23, 13.20 to 14.88 m), B2b (SD 64 NE/7, 10.40 to 19.53 m; SD 64 NE/19, 22.70 to 22.80 m; SD 64 NE/20, 63.70 m and SD 64 NE/23, 6.10 to 9.60 m) and P2c (SD 64 NE/17, 26.02 to 30.00 m). The junction with the Upper Bowland Shale Formation was seen in boreholes SD 64 NE/17 and 20.

Throstle Nest Anticline

The maximum thickness of the formation in this anticline occurs in the south-west, where the formation is 360 m thick, but it thins dramatically north-east by diminution of the Park Style Limestone and Pendleside Sandstones members. Along the north-west limb these members give rise to prominent bench and scarp features, which trend north-north-east. The middle leaf of the Pendleside Sandstones is very conspicuous and its prominent feature is broken by the deeply dissected valleys of Leagram Brook [6260 4555], Rathera Clough [6290 4622], Hell Clough [6340 4685] and Dinkling Green Brook [6380 4785]. The crest of the feature is drift free, and several small quarries reveal minor sections. Only Leagram Brook affords a good section; the valleys of the other streams mentioned above are choked with head and till.

In Leagram Brook [6298 4511 to 6260 4561], near Park Style, there is nearly continuous exposure from the base of the formation up to the base of the upper leaf of the Pendleside Sandstones Member (Figure 8). This is also the type section for the Park Style Limestone Member, described earlier. Several ammonoid bands have been located in the lower part of the formation including the Goniatites hudsoni and G. globostriatus bands, and the "Dimorphoceras band" of Moore (1939), which hitherto was known only from 6.09 m below the Goniatites crenistria band at Dinckley Ferry (Earp et al., 1961) in the adjacent Clitheroe district. *Goniatites granosus* was found 5 m above the middle leaf of the Pendleside Sandstones. The base of the upper leaf is exposed some 2.7 m higher in the section. This leaf is about 32 m thick, but dies out rapidly north-east. A parallel, but less well-exposed sequence to that seen in Leagram Brook occurs in Hell Clough [6355 4633 to 6350 4655], some 1.3 km to the north-east. The base of the formation is not exposed, but *Goniatites hudsoni* has been found in loose "bullions" in the stream and *G.* cf. *globostriatus* was located 3 m below the base of the Park Style Limestone Member. The sequence in this member at Hell Clough differs from that in the type section in that only the higher wackestone-dominated part is present. The member is only 6.3 m thick and is too thin to be mapped north-east of Hell Clough. The interval between the Park Style Limestone Member and the lowest leaf of the Pendleside Sandstones Member comprises black mudstones and siltstones; as at Leagram Brook no stratigraphically significant fauna was obtained. The lower leaf of the Pendleside Sandstones is 4 m thick (compared to 8 m at Leagram Brook) but poorly exposed, as are the intervening 53 m of mudstones and siltstones between the lower and middle leaves.

A small unexposed inlier of the formation is inferred from geometrical considerations around Dairy Barn [6328 4353], on the south-eastern limb of the anticline.

Chipping Anticline

Exposure of the Lower Bowland Shale Formation is poor in the Chipping Anticline. On the north-western limb, beds near the base of the Park Style Limestone Member are exposed in Chipping Brook [6216 4349], opposite the chair works. The section comprises 5 m of well-bedded, dark grey wackestones and packstones, with sharp erosive bases and graded beds, interbedded with black mudstone. Poorly preserved fragments of *Goniatites* ex gr. *globostriatus* are present. In a ditch 540 m north-west of Dairy Barn [6279 4378], a similar 0.7 m-thick sequence of graded packstones and black mudstones yielded a bullion containing a prolific B2b Ammonoid Subzone fauna, including *Beyrichoceras* cf. *obtusum*, *Beyrichoceras* sp. ex gr. *rectangularum/vesiculiferum*, *Bollandites phillipsi*, *Bollandoceras micronotum*, *Dimorphoceras* cf. *gilbertsoni*, *Goniatites moorei*, *Metadimorphoceras pseudodiscrepans* and *Nomismoceras vittiger*, associated with productoid and chonetoid brachiopods, bivalves and the trilobite *Metaphillipsia seminiferus*. Although this exposure undoubtedly lies in strata equivalent to the lower part of the Park Style Limestone Member, there is not enough evidence to infer the nature and extent of the outcrop beneath the drift to the south of the exposure.

Saunders Anticline

On the north-western limb, beds low in the formation are best exposed in Chipping Brook [6140 4381 to 6135 4388], near Nan Kings Farm (Figure 8); they lie within the Park Style Limestone Member. The lower packstone-rich division is 21.7 m thick and shows graded beds with some slumping. Poorly preserved ammonoids obtained from black mudstones exposed immediately north of the bridge leading to the disused Saunder Rake Works, include *Goniatites* sp. with crenistriate ornament (possibly *G. hudsoni* or *G. crenistria*, representing a B2a or P1a Ammonoid Subzone horizon respectively). The lower 10.1 m of the upper division is predominantly mudstone and this is overlain by 8.49 m of grey to dark grey, massive wackestone which, because of selective weathering of folded and podded chert laminae, shows conspicuous internal slump folds. This bed was also located in Leagram Brook (Figure 8; slumped nodule bed).

Black mudstones immediately underlying these wackestones, exposed in a meander cliff on the right bank, yielded a crushed fauna including *Goniatites* cf. *striatus*, *Nomismoceras vittiger* and *Pronorites* sp., suggesting a P1b Ammonoid Subzone horizon.

The middle leaf of the Pendleside Sandstones Member forms a conspicuous, but largely drift covered, south-west-trending bench feature on the north-west limb. Two disused quarries on Lingey Hill [6067 4348 and 6069 4358] provide the best exposure; the northernmost shows a 3 m section of pale grey to yellow-brown, fine- to medium-grained sandstone in massive, composite beds.

On the south-eastern limb a good section in the basal part of the formation is present in Dobson's Brook [6192 4407 to 6201 4416] (Figure 8), deeply incised between Nan King's Farm and Leagram Hall Farm; the base of the section exposes the Goniatites hudsoni band. The overlying *G. globostriatus* band was not located, but *G. globostriatus* was recovered from a 1.5 m-thick slumped packstone 6.77 m above the base of the section. A further 6 m of poorly exposed black mudstones with thin packstones are seen upstream, overlain by 6 m of thick-bedded, sharp-based, graded packstones and interbedded mudstones representing the lower division of the Park Style Limestone Member. Loose blocks from near the base of the member yielded a rich ammonoid fauna, including *Bollandoceras globosum*, suggestive of a high horizon within the B2b Ammonoid Subzone.

Blacksticks Anticline

The Park Style Limestone Member forms the core of the anticline in the east. The best exposure is at the disused and now largely in-filled Blackhall Quarry [6108 4307]. This quarry provided many of the ammonoids described by Phillips (1836). The south-west face and adjoining stream provide the only section of what was once an extensive excavation trending for 300 m north-east along strike. The exposure, first recorded by Hind and Howe (1901), comprises 12 m of graded, sharp-based packstones, interbedded with mudstone; the packstones yielded a juvenile *Beyrichoceras*. The beds dip steeply south-east and are probably on the north-western limb of a minor syncline, the axis of which is detectable in a stream to the south-west [6096 4296].

Further west, near Richmond Houses, a north-west-trending fault throws down to the south-west, causing the Pendleside Sandstones Member to crop out in the anticlinal core. This is exposed in the gorge of the River Loud, known as Hough Clough [5938 4256 to 5953 4244], about 400 m east-north-east of Blacksticks Farm. The succession is estimated to be 186 m thick, measured on the north-western limb, and is assumed to be virtually the full thickness of the member, but only half of this is actually exposed. The south-eastern limb of the anticline, exposed downstream [to 5979 4215], has more exposure, but is also structurally complex, with several reversals of dip direction and minor faulting.

The sandstones are thinly to thickly bedded, grey (yellow-brown when weathered), fine-grained and weakly calcareous. The beds are sharp based, normally graded and mostly lack internal lamination, although a few beds show some parallel lamination; sole structures are rarely seen. Some of the beds show a thin basal bioclastic division in which the bioclasts have been removed by solution, leaving moulds. Small mudstone flakes and carbonaceous plant fragments also occur. Varying proportions of interbedded, dark grey, silty mudstone are normally present and locally constitute the dominant lithology, accompanied by dark, argillaceous graded packstone beds.

Away from the core, the north-western limb of the anticline is gently folded, giving a broad outcrop of the Pendleside Sandstones Member. Good exposure is seen in a disused roadside quarry [5816 4372], 600 m west-south-west of Lower Core, where some 8.0 m of grey, fine-grained sandstone, similar to those in the upper part of the sequence at Hough Clough, are exposed (Plate 8). The beds are up to 1.6 m thick and massive, although some display lamination near the top; shaly intercalations and sole marks are scarce. Another disused quarry [5843 4417], north-west of Lower Core, exposes 1.6 m of similar sandstones.

Beds overlying the Pendleside Sandstones Member are exposed in several stream sections on the north-western limb of the anticline. A 6 m sequence of dark, thin-bedded, argillaceous packstones and black mudstones is exposed in a stream [5837 4370] 400 m south-west of Lower Core. Dips are variable due to the proximity of a

Plate 8 Pendleside Sandstones in a disused quarry [5816 4372] near Lower Core Farm (A14826).

branch of the Oakenclough Fault, which separates these beds from the main outcrop of Pendleside Sandstones. The ammonoid *Goniatites granosus* was found, indicating a P2a Ammonoid Subzone horizon. A 3.3 m section is present in a stream at Woodgates [5963 4390], comprising tough calcareous siltstone and argillaceous wackestone, with a few black chert lenses and shaly beds. Similar beds are poorly exposed in a small disused quarry [5930 4416] at Higher Core, where they may have been worked for hardcore used in local road making.

On the south-eastern limb, beds lying above the Pendleside Sandstones are seen only in scattered exposures in a stream east-north-east of Hall Trees Farm. One exposure [6071 4220] shows a 0.28 m graded bed of crinoidal packstone overlain conformably by bullions containing *Cravenoceras leion*, marking the base of the Upper Bowland Shale Formation.

Thornley Anticline

The outcrop in the west is largely conjectural due to lack of exposure, but on the flanks of the Vale of Chipping, in the eastern part of the anticline, exposure is moderately good.

The north-western limb is complicated by tight minor folding adjacent to the Doeford Fault. A disused and partly backfilled quarry at Wallclough [6042 4152] exposes about 11 m of vertically dipping, dark argillaceous wackestones, with thin beds of black mudstone assigned to the upper part of the Park Style Limestone Member. The mudstones contain *Posidonia becheri*, suggestive of an interval within the P1b–c ammonoid subzones. Beds above the Park Style Limestone Member are exposed in a stream by the Old Vicarage, near Astley House [6091 4198 to 6086 4202], revealing a structurally complex section in thinly interbedded cherty limestone and mudstone. In another stream section, upstream from Astley House [6095 4209], a steep-limbed, minor synclinal structure plunging south-west is cored by the Pendleside Sandstones Member.

The south-eastern limb displays numerous discontinuous exposures in ditches and streams on the north-western slopes of Longridge Fell. A relatively good section occurs in a ditch [6377 4090 to 6392 4081] south of New House, near Thornley Hall, where 120 m of the sequence below the Pendleside Sandstones is discontinuously exposed. Dark grey, subfissile mudstones predominate, but no marker bands were noted. However, between about 19 m and 31 m above the base of the section a sequence of thinly bedded, commonly graded packstones and wackestones is present, and this may correlate with the lower part of the Park Style Limestone Member. Similar limestones exposed in a stream [6175 3938] 250 m west of White Fold, about 80 m above the base of the formation, yield well preserved uncrushed ammonoids in the base of a graded packstone bed, including *Goniatites* cf. *warslowensis, Nomismoceras* sp. and *Pronorites ludfordi*, associated with a hash of fragmentary orthocones, productoids, crinoids and ostracods. This assemblage indicates an horizon low in the P1c Ammonoid Subzone. Another exposure upstream [6185 3930], some 67 m higher in the sequence, shows black mudstones from which *Hibernicoceras* sp. and *Posidonia becheri* were found, suggestive of a level within the P1c–d ammonoid subzones.

The Pendleside Sandstones Member is exposed in several streams along Longridge Fell. One section [6451 4103], 300 m east-north-east of Brook House, shows 14.3 m of thick-bedded sandstone with shaly intercalations, which are caught up in a tight anticline and syncline. These beds lie in the middle of the member. The basal 10.5 m of the member is exposed in a ditch nearby, on the north-east side of the road to Jeffrey Hill, where blocky sandstones are interbedded with silty and sandy mudstones, especially in the lower part of the section. Farther south-west, between Dale House [6285 3996] and Birks [6165 3907], several gullies dissect the bench feature formed by the sandstones. Although only a single leaf is mapped, thin (0.3 m) sandstone beds occur [6263 3988; 6265 3988] up to 8 m below its base. In the stream section [6285 3972 to 6273 3982] at Mercers Wood, south-west of Dale House, some 80 m of sandstone overlie these beds. The principal lithology is greyish green, fine- to medium-grained sandstone. Many beds are flaggy and micaceous, and some show ripple marks on bedding surfaces. Crinoid ossicles are present at several horizons, but are particularly prevalent in the lower beds, which are very calcareous. Intercalations of mudstone occur throughout, but are subordinate to sandstone.

At Mercers Wood, beds between the top of the Pendleside Sandstones Member and the top of the formation, comprise 24 m of fissile mudstones and platy calcareous mudstones, some with ammonoids. A similar section is seen in the stream above White Fold [6217 3935], where a bullion band overlying the Pendleside Sandstones Member yielded well-preserved ammonoids, including *Goniatites granosus* (Plate 1j,k) and *Lusitanites subcircularis*, indicating a high horizon in the P2a Ammonoid Subzone. Some 17.0 m higher in the succession [6217 3934], an exposure of calcareous platy mudstone yielded a fauna including *Lyrogoniatites georgiensis*, the index ammonoid of the P2c Ammonoid Subzone, marking the top of the formation. A small hillside exposure [6274 3984] 200 m south-west of Dale House yielded bullions containing *Sudeticeras splendens*, indicative of the P2b Ammonoid Subzone. A parallel section to that exposed around Mercers Wood is seen to the north-east, in a ditch 380 m east of Brook House [6458 4093].

In the far west, near Barton, borehole SD 53 NW/4 penetrated 8.54 m of mudstone assigned to the Lower Bowland Shale Formation, below the base of the Manchester Marls, which was at a depth of 67.36 m. The sequence comprises 5.94 m of red-stained and grey mudstone with ironstone bands, overlying 2.6 m of black mudstone containing *Hibernicoceras* sp. and *Posidonia becheri*, suggesting a probable P1c Ammonoid Subzone correlation and demonstrating the continuation of the Thornley Anticline beneath the Permo-Triassic cover.

Slaidburn Anticline

Much of the outcrop in this anticline lies in adjacent districts. In the present district the outcrop on the north-west limb is unexposed and therefore conjectural; the Pendleside Sandstones Member is believed to be absent. The south-eastern limb is truncated by the north-west-trending Giddy Bridge Fault, which has thrown the Lower Bowland Shale Formation down below the surface, except for a small faulted inlier at Giddy Bridge Plantation [6825 4937]. Here black mudstones exposed in a meander cliff on the right bank of Birkett Brook contain *Hibernicoceras* sp. and *Posidonia becheri*, favouring an horizon within the P1c or P1d ammonoid subzones.

NJR

THREE

Namurian rocks

Rocks of Namurian age occur in the eastern half of the district in two main areas separated by the Dinantian outcrop in the Vale of Chipping (Figure 9). In the larger northern outcrop, the Upper Bowland Shales and the Pendle Grit are the thickest, most extensive and well-exposed formations, with younger Namurian strata brought into the east and north of Garstang by a regional plunge to the west. Ex-

posure here is generally poor, however, and the sequence is little known in detail, with few stratigraphically significant exposures.

The south-eastern outcrop is disposed around a broad synclinal structure, the Ribchester Syncline (Figure 21). The Pendle Grit is the most prominent and well-exposed formation, but here, unlike the northern outcrop, it is overlain by

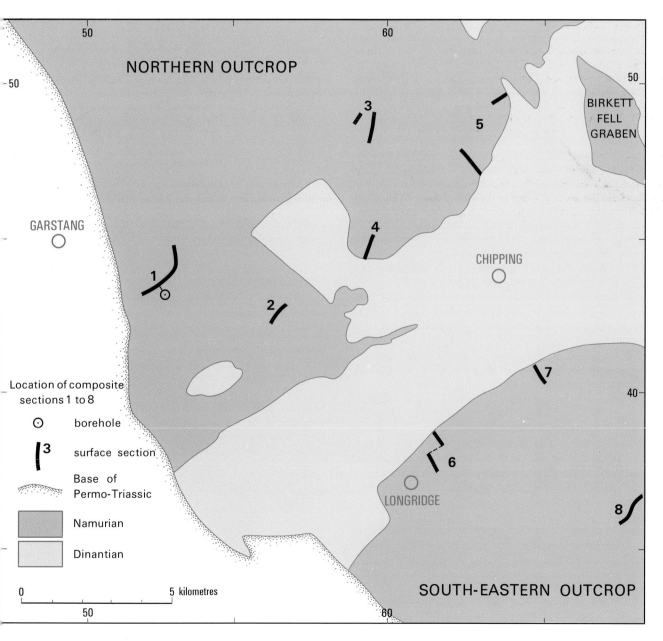

Figure 9   Outcrops of Namurian rocks and location of numbered composite stratigraphical sections shown in Figure 11.

another major sandstone, the Warley Wise Grit. Much of the outcrop of the latter formation, like that of the overlying Sabden Shales, is conjectured beneath thick drift.

PALAEOGEOGRAPHY AND DEPOSITIONAL HISTORY

In early Namurian (Pendleian, E1) times the Garstang district continued to lie within the Craven Basin. This basin of subaqueous, mainly marine deposition had developed by crustal extension during the previous epoch (see p.4) and was continuing to undergo thermal subsidence (Leeder, 1988) together with the adjacent shelf areas. The margins of the basin, marked especially by the Craven Fault Zone at the edge of the Askrigg Block to the north-east (Arthurton et al., 1988), were also well submerged, so that the supply of coarser sediment from the extrabasinal shelves and hinterlands had been cut off.

The early Pendleian sediments were largely dark, thinly intercalated, weakly calcareous, muddy silts and hemipelagic muds laid down mainly from suspension in the relatively deep, poorly oxygenated waters. Sparse bivalve and fish faunas indicate a restricted marine or nonmarine environment. Occasional beds with a more abundant and varied fauna, including thick-shelled ammonoids (goniatites), indicate fully marine conditions. These 'marine bands', generally more calcareous and less silty than the rest of the muddy sequence, mark the sea-level highstands of minor transgressive cycles (Ramsbottom et al., 1962; Holdsworth and Collinson, 1988) superimposed on the general subsidence. Eustatic sea-level changes, caused by glacial fluctuations in the southern hemisphere (Powell and Veevers, 1987) now seem to be generally accepted as a cause of this minor cyclicity (Leeder, 1988), which affected sedimentation throughout Namurian and Westphalian times.

At least three minor cycles are recognised in the Pendleian sequence in this district, marked at their bases by the Cravenoceras leion, Eumorphoceras pseudobilingue and Cravenoceras malhamense marine bands; they all lie in the Upper Bowland Shale Formation. Immediately before the C. malhamense marine transgression there was a minor influx of calcareous feldspathic sand (Hind Sandstone) into the district, deposited by turbidity currents in a few localised channels on the basin floor. This was the local precursor of a huge influx of coarse feldspathic sand that followed this marine transgression and formed the Pendle, Warley Wise, Grassington and Brennand grits. The influx of sand ended when a further eustatic transgression took place at the beginning of Arnsbergian times, represented by the Cravenoceras cowlingense Marine Band.

The sand can only have been transported to the Craven Basin by a great river system, probably on the scale of the present day Brahmaputra. According to Sims (1988), transport was from the north or north-east and the sands were first deposited from suspension in channels on the basin floor, followed by the advance of the delta front across the district. The river system itself only reached the edge of the district, for evidence of fluviatile facies is restricted to the Warley Wise Grit in the highest part of the sequence in the Ribble valley.

The cause of the dramatic influx of sediment is not yet resolved. Sims (1988) argued that there is no need to invoke extraordinary events, such as massive tectonic uplift in the source areas, since feldspathic sand had been intermittently deposited in other parts of northern Britain from early Dinantian times (see, for example, Sevastopulo, 1981). Instead, Sims favours geomorphological changes such as river avulsion or river capture to account for deltaic deposition switching into the 'Bowland and Lancaster Fells basins' (Craven Basin). Other factors, such as climatic changes consequent on the convergence of the Gondwana and Euramerican landmasses in equatorial latitudes (Rowley et al., 1985), may also have had an influence.

After the brief progradation of fluvial conditions into the eastern margin of the district near the end of Pendleian times, a deeper-water environment appears to have again prevailed for most of the Arnsbergian (E2). At times, however, a southwards-prograding delta floodplain approached the northern boundary of the district. This is indicated by thin coals and seatearths at several levels in the early Arnsbergian sequence in the Lancaster district to the north (Wilson et al., 1989; A Brandon, personal communication). Exposure and biostratigraphical control are poor, however, and these conclusions are only tentative.

During deposition of the Caton Shales (in E2b times) one or more sea-level rises deepened and extended the basin. Evidence from the Lancaster district (Wilson et al., 1989) suggests that a similar process occurred in late Arnsbergian and early Chokierian (H1) times during deposition of the Crossdale Mudstones. Later in the Chokierian, intermittent fluviodeltaic conditions probably became more prevalent and the cross-bedded Wellington Crag and Ellel Crag sandstones were deposited.

CLASSIFICATION

Lithostratigraphy

In keeping with modern practice, most of the major rock or lithostratigraphical units that have been delineated on the new geological maps of the district have been assigned formal names in accordance with the heirarchical scheme set out by Holland et al. (1978). Although there is no formally defined name for the arenaceous rocks of Namurian age in the region, 'Millstone Grit' continues to be used as a convenient informal term. Originally, it was used to denote the dominant coarse-grained Carboniferous sandstone facies of the Pennines, but subsequently it acquired a chronostratigraphical connotation when used in the form 'Millstone Grit Series'. This hybrid term has become less used in recent years because it inappropriately includes substantial thicknesses of argillaceous rocks in its lower part, especially in the southern Pennines. In the present district 'Millstone Grit Group' is used to include all strata of Namurian age above the base of the Pendle Grit Formation, a usage similar to that adopted in the adjacent Clitheroe district (Earp et al., 1961).

Units now assigned formational status are the Upper Bowland Shales, the Pendle Grit, the Warley Wise Grit (and the equivalent Brennand Grit), the Roeburndale Formation, the Caton Shales, the Heversham House Sandstone, the

Crossdale Mudstone, the Ellel Crag Sandstone and the Sabden Shales (Figure 10). Other units with less extensive outcrops are given member status: they are the Copster Green Sandstone, the Park Wood Sandstone and the Stonehead Sandstone. All other lithostratigraphical names, including those of marine bands, are used informally. Details of stratotype sections situated outside the district and not given in the text may be found in BGS archives.

The first Geological Survey map of the district, published in 1883, delineated 'Bowland Shales' and 'Pendle Grit'. These units are now formally defined as a group and a formation respectively (see above). The names used for the overlying sandstones, i.e. 'Fourth (Kinderscout) Grit' and 'Third Grit', subsequently proved neither useful nor correct

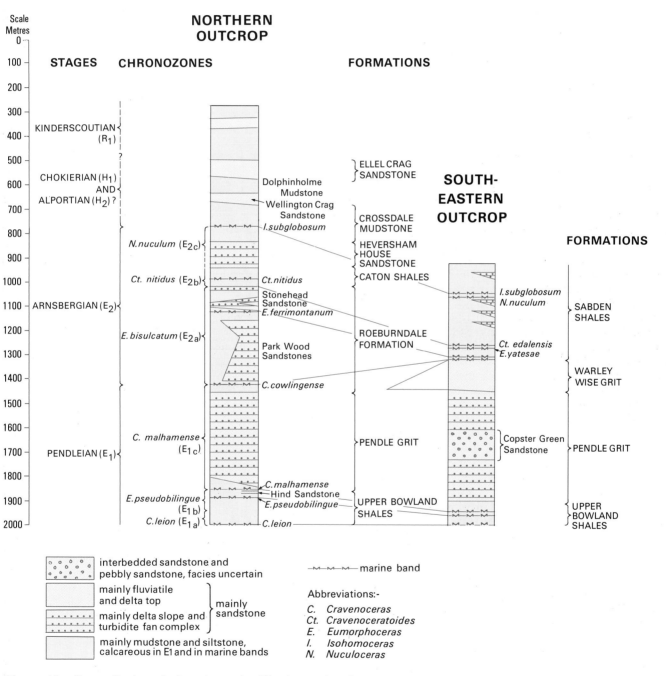

Figure 10 Generalised vertical sections, classification and major sandstone facies of Namurian rocks in the district.

in their implied correlation with similarly named sandstones elsewhere in the Pennines, and they are no longer used in this district.

Chronostratigraphy and biostratigraphy

The Namurian Epoch, representing the interval of time when these strata were being deposited, is divided into stages and chronozones (Ramsbottom et al., 1978). The boundaries of these chronostratigraphical subdivisions (Figure 10) are recognised biostratigraphically by the presence of diagnostic ammonoid (goniatite) faunas. Ammonoids are free-swimming marine animals that mostly evolved rapidly during Namurian times; thus their fossilised remains provide a ready means of widespread correlation of the mudstones in which they are preserved. Where ammonoids have not been recovered, it has usually been possible to find other fossils, such as bivalves and miospores, that indicate the likely age of the strata, or to use combinations of more subtle lithological and biostratigraphical evidence.

It should be noted that the name of the ammonoid *Eumorphoceras pseudobilingue* has been changed to *Tumulites pseudobilinguis*, but that the original name Eumorphoceras pseudobilingue Marine Band has been retained on the geological maps of the district.

BOWLAND SHALE GROUP

Upper Bowland Shale Formation

The term Upper Bowland Shales first appeared on a diagram by Bisat (1928, pl. vi). It was used by Stephens et al. (1953), by Ramsbottom (*in* Rayner and Hemingway, 1974, 76–78) and most recently by Arthurton et al. (1988) in the adjacent Settle district. The name is now assigned formational status, using the definition given by Arthurton et al. (1988, p.72), and refers to the mudstone sequence between the base of the Cravenocerous leion Marine Band and the base of the Pendle Grit Formation (Figure 11). This sequence forms the uppermost part of the section in Little Mearley Clough (Earp et al., 1961, fig.8) in the adjacent Clitheroe district which is proposed as the type section for the formation. The base and the top are conformable in both districts.

Only one part of the formation is here formally distinguished as a member, namely the Hind Sandstone, which was first mapped and named by Moseley (1962). It occurs between the Eumorphoceras pseudobilingue and Cravenoceras malhamense marine bands and is therefore of E1b age.

In the Garstang district, the Upper Bowland Shales form an extensive though largely drift-covered outcrop between the boundary with the Permo-Triassic rocks and the Oakenclough Fault. Elsewhere in the district, the outcrop mostly forms a narrow belt below the major escarpment of Pendle Grit, both around the southern margin of the Bowland Fells and the north side of Longridge Fell. Irregular outcrops of small extent lie in the Vale of Chipping, to the west of Chipping, and are also associated with the Birkett Fell Graben in the east. There are no significant borehole provings of the Upper Bowland Shales in the extensive areas where younger strata crop out.

Thicknesses are greatest along the Bowland Fells escarpment, where a range between 185 m and 225 m is estimated, and least along the Longridge Fell outcrop, where estimates range from 67 to 115 m. Variations in thickness are shown graphically in Figure 11.

Lithologically, the greater part of the formation consists of thinly interbedded, dark, fissile mudstone and weakly calcareous and dolomitic, blocky or platy silty mudstone and siltstone. In some parts of the sequence the carbonate content is higher, to the extent that locally mappable argillaceous limestones and dolostones occur (Howard, in preparation), with individual beds up to 0.70 m thick. The quartz silt in these blocky beds is mostly finely disseminated but may also be concentrated in fine pale laminae at or near the bases of some beds. Slight normal grading probably indicates deposition from weak turbidity currents. Some very fine mica and finely comminuted debris of carbonaceous plants and calcareous bioclasts may also be present. The blocky siltstones and argillaceous limestones are hard and tough in unweathered sections such as that present in the gorge of the River Brock below Walmsley Bridge. More commonly, however, weathering has decalcified and softened them, to produce a yellow-brown colouration (Plate 9) and a finely porous texture. These rocks generally form parallel beds of even thickness, which are varied, particularly in marine bands, by a few concretionary swellings or bullions, and by the presence of very thin siliceous limestones or chert lenses. When freshly broken, the blocky beds commonly emit a bituminous odour, and small oil bleeds are occasionally observed.

In the marine bands, macrofossils are commonly abundant, but the fauna is of low diversity, with posidoniid bivalves, ammonoids and nautiloids being the principal constituents. Well-preserved uncrushed shells occur in some bullions, especially in the basal parts, but in the fissile mudstone intercalations most fossils have been crushed by compaction. By contrast, macrofossils are not common in the thick sequences between marine bands; they are limited to small posidoniid bivalves, spat, fish and plant fragments. Bioturbation is commonly present but individual trace fossils are very diffuse and ill defined (Howard, in preparation).

Sequences of noncalcareous barren mudstones and siltstones with a few thin sideritic ironstone nodules or beds occur in the higher parts of the formation, above the E. pseudobilingue Marine Band. The Hind Sandstone is known from only a few scattered localities (see below) and is absent in most well-exposed sections. The sandstone is generally grey when fresh, weathering to rusty brown, and is medium to coarse grained and thick bedded. The beds lack grading and internal lamination, but are probably turbidites. The general lateral impersistence of the member is well exemplified in one of the best exposures, at Blue Scar on Mellor Knoll [6570 5456] in the adjacent Lancaster district, by visible lateral impersistence of individual beds. These features indicate that the sand body represented by the member has a channel form, but the orientation of the channel (or channels) has not been determined. In the highest sequence, above the C. malhamense Marine Band, there is a gradational coarsening upwards, heralding the incoming of the Pendle Grit Formation.

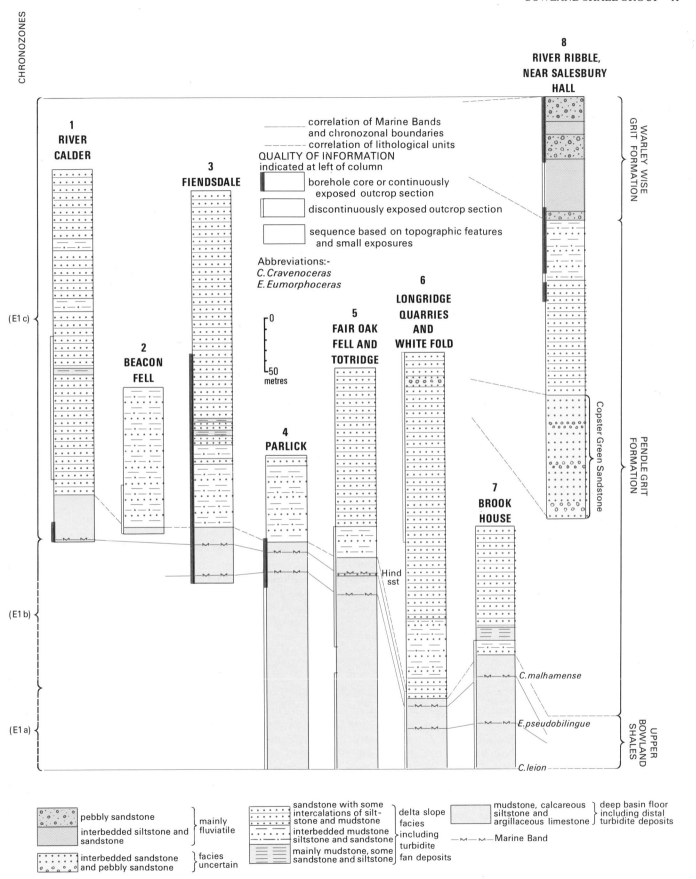

Figure 11 Composite sections of strata of early Namurian (Pendleian) age (see Figure 9 for location of sections).

Plate 9 Excavation in Upper Bowland Shales near Blindhurst [5888 4504] exposing a typical interbedded sequence of blocky, yellow-brown weathering, calcareous mudstones and siltstones, and dark grey fissile mudstones (A14794).

Details

Claughton, Inglewhite and Beacon Fell

The Cravenoceras leion Marine Band is exposed at two places near Inglewhite, exposures that help to fix the otherwise largely conjectural position of the boundary between rocks of Dinantian and Namurian ages in an area where the strata are strongly folded and largely drift covered. The first locality [5364 4078], north-east of Lower House, is in a stream flanking the north-west side of the anticlinal inlier of Lower Bowland Shales around Scotch Green [541 405]. Here, some 5.70 m of calcareous silty mudstone with fissile mudstone intercalations have yielded a sparse fauna of poorly preserved bivalves and ammonoids in the basal 1.67 m, including one specimen of *C. cf leion*. The second locality [5455 3903] is in a discontinuous 120.93 m-thick section exposed in Sparting Brook, including 103.17 m of strata assigned to the Upper Bowland Shales; full details are given by Bridge (1988c). The beds likely to belong to the *C. leion* Marine Band comprise of 3.3 m of dark grey, blocky, calcareous siltstone with limestone bullion bands, from which poorly preserved bivalves and ammonoids including *Cravenoceras* sp. have been extracted. No other significant fossils have been found in the Upper Bowland Shales part of this section, probably because nearly half the sequence is obscured. Siltstone with sandstone laminae, occurring in the uppermost 6.9 m of the section, probably correlate with similar beds in the middle part of a discontinuous section with a total thickness of about 60 m in Factory Brook, a higher tributary stream [5509 3940 to 5507 3956] near Inglewhite (Bridge, 1988b).

The River Brock, from a point [5279 4129] below Walmsley Bridge to Brock Bottom [5469 4225], has, in places, incised through the thick till into the underlying Upper Bowland Shales. This part of the river's course roughly follows the strike of the outcrop at around the level of the *E. pseudobilingue* Marine Band. This band is partly exposed at a bend in the river [5372 4166] just above water level, 0.90 to 1.15 m above the base of a 28 m section of calcareous siltstones with shaly mudstone intercalations. The sparse fauna includes *Posidonia trapezoedra* and *Tumulites* cf. *pseudobilinguis*. The best section showing the unweathered lithology, though without marker bands, is in the gorge [5315 4129 to 5327 4140] below Walmsley Bridge. A total of 9.5 m are exposed here, consisting of barren, tough, blocky, medium to dark grey, weakly calcareous siltstone and dolostone and thinly interbedded, darker, fissile mudstone in a ratio of about 10 to 1. The sequence shows characteristics of fine-grained turbidites, including normally graded, sharp-based beds. Many of these have silt or fine sand basal laminae; they are similar either to the DE or CDE type beds of Bouma (1962) or to those more recently described, for example, by Stow and Shanmugam (1980). A few beds show cross-lamination and sole structures, including load casts and rare flute casts. Palaeocurrent readings from two flute casts and three sets of cross-lamination suggest a flow direction generally towards the east. A thin section (E 61941) of one sample of carbonate siltstone, examined by B Humphreys, shows it to consist of ?sponge spicules set in a matrix of calcite and dolomite that forms over 50 per cent of the rock.

To the south-east, a little gorge with a waterfall [5634 4197], near Crombleholme Fold, exposes some 8.0 m of calcareous siltstone in beds up to 0.73 m thick, with shaly intercalations; some of the latter are rich in small, poorly preserved *P. corrugata*. A few 'ghost' ammonoids are also present, which indicate that the stratigraphical level of the section is probably at or near the E. pseudobilingue Marine Band. The rock faces of this gorge may have been quarried for road ballast, since they look too fresh and extensive to have been produced by natural agencies.

An exposure in a large ditch [5289 4292] near Crosshouse, Claughton, of about 3.0 m of rather soft, brown, medium-grained sandstone with shaly intercalations is assigned to the Hind Sandstone. Low dips here suggest that the sandstone lies on an anticlinal axis. Its correlation with the Hind Sandstone is indicated by the likely geometrical relationship with the C. malhamense Marine Band proved some 590 m to the north-north-east in a borehole [SD 54 SW/17] between depths of 42.67 and 44.20 m. The borehole terminated at 46.33 m just at or above the depth where the Hind Sandstone would have been expected.

The only other occurrence of the C. malhamense Marine Band in this area is in a stream [5679 4219] at Woodfold on the southern slope of Beacon Fell. Here a fauna comprising *P. corrugata*, *P. membranacea* and *C. malhamense* was extracted from poorly exposed, dark, fissile mudstone intercalations between flaggy, calcareous siltstone beds up to 0.25 m thick. These beds appear to lie with some discordance on 1.6 m of massive, medium- to coarse-grained sandstone, probably a local development of the Hind Sandstone. The outcrop of this member has not been traced away from the exposure due to lack of evidence.

Outcrops west of Chipping

Two small synclinal outcrops of the Upper Bowland Shales occur west of Chipping. The presence of the first outcrop [587 430], lying between Watergate Farm and Brown Brook Farm, is inferred from the general structure of the area; there are no exposures or borehole provings. The second outcrop [604 420] lies immediately south-east of Hall Trees Farm where it is bounded on its southern side by the Doeford Fault. The presence of Upper Bowland Shales is proved by an exposure of the C. leion Marine Band in a steep bank [6071 4220] east of Hall Trees Farm. The well preserved *C. leion* fauna was extracted from a 0.07 m limestone bullion, overlain by about 15 m of discontinuously exposed calcareous siltstones and argillaceous limestones with shaly mudstone intercalations.

Bowland Fells escarpment and Sykes

The south-eastern escarpment of the Bowland Fells follows an indented course between Bleasdale in the south-west and Staple Oak Fell in the north-east. The outcrop of Upper Bowland Shales forms a meandering belt on the face of this escarpment, with stream gullies providing numerous sections, especially on Parlick, Fair Oak Fell and Totridge (Figure 9, and Figure 11, columns 4 and 5). The escarpment also encircles the Dinantian anticlinal inlier at Sykes, but the main exposures here lie in the adjacent Lancaster district (see Hughes, 1986). Additionally, a small inlier is present in the valley of the Fiendsdale Water in the heart of the Bowland Fells [598 494].

The Cravenocerous leion Marine Band is exposed at only one locality in the area [5924 4437], a small stream 200 m north of Higher Core. The exposure comprises 1.37 m of dark shaly mudstone interbedded with argillaceous limestone, one bed of which thickens into a 0.15 m bullion yielding solid goniatites including *C. leion*. The overlying beds, probably of E1a age, are discontinuously exposed in nearby ditches [5907 4438; 5906 4451] and consist of calcareous siltstones and argillaceous limestone beds up to 0.7 m thick, with sporadically fossiliferous interbedded mudstones. Similar beds are exposed in several of the gullies on the north-east slopes of Totridge (Howard, in preparation fig. 2). At a higher level in the sequence the argillaceous limestone beds become fewer and thinner, with a corresponding increase in the proportion of fissile mudstones. A typical sequence in a small quarry north of Blindhurst [5888 4504] is shown in Plate 9.

The C. brandoni Marine Band (E1b1) has not been found in the district but may be represented by a fauna of *P. corrugata* and indeterminate ammonoids collected from a stream [6322 4741] below Fair Oak Fell, one of several there with good exposures of the middle and the middle and upper parts of the formation. The C. brandoni Marine Band is exposed just north of the district boundary on the north-west side of Staple Oak Fell in the Sykes Anticline, where it lies 60 m above the C. leion Marine Band (Hughes, 1986). Mappable, argillaceous, limestone-rich sequences occur on the north-east slopes of Totridge not far below the mapped outcrop of the E. pseudobilingue (E1b2) Marine Band. Individual limestone beds range in thickness from 0.08 to 0.50 m and collectively form limestone-with-mudstone packages 5 to 10 m thick (Howard, in preparation). The upper part of the formation is well exposed in a number of places, with both the E. pseudobilingue (E1b2) and C. malhamense (E1c) marine bands present. These localities are on the right bank of Fiendsdale Water [5963 4935], west of Bleadale Ridge, and in two gullies [5933 4537 to 5940 4581; 5919 4499 to 5929 4496] on the west face of Parlick. Other exposures of the E. pseudobilingue Marine Band occur nearby [5962 4882; 5956 4570], along the escarpment to the north-east [6345 4826] near Whitmore Farm and on Mellor Knoll [6500 4970]. The thickest development of the E1b2 band is in the 37.55 m Fiendsdale Water section, where the band comprises:

	Thickness m
Mudstone, calcareous, friable with *Tumulites pseudobilinguis*	1.90
Limestone, argillaceous	0.30
Limestone, nodular, argillaceous with *T. pseudobilinguis*	0.35

P. corrugata and mollusc spat occur in the 4.0 m of mudstones and thin argillaceous limestones immediately overlying the above sequence. The thickness of strata between the two marine bands is much greater in the Fiendsdale Water section (26 m) than in the two Parlick sections (15.00 m and 16.8 m) 4 km to the south. Details of these sections are given by Howard (in preparation) and Aitkenhead (1990b). The Hind Sandstone, present in this interval in some other parts of the district, is known in this area at only one locality [6293 4739], below Fair Oak Fell, where it is 1.2 m thick and lies 0.4 m below the C. malhamense Marine Band (see below). The sandstone is described by Howard (in preparation) as a structureless, coarse-grained, feldspathic greywacke with common lithic fragments. Both sorting and grading are poor and there is a considerable admixture of silt and clay-grade sediment, together with conspicuous mica grains and abundant small carbonaceous fragments. This description is also representative of the member elsewhere.

The exposed sequences above the C. malhamense Marine Band vary in thickness from 6.5 to 8.65 m and show a gradational coarsening-upwards transition into the thinly interbedded silty mudstones and turbiditic muddy sandstones of the basal part of the Pendle Grit Formation.

Birkett Fell graben and Cow Ark

The Upper Bowland Shales in this area are best exposed in Crimpton Brook [6838 4678 to 6819 4712], in a section of thinly interbedded blocky and fissile mudstones, which are pyritic and ferruginous in places. The only marker band present is that of *T. pseudobilinguis* (*E. pseudobilingue*), which is associated with large limestone bullions and shelly, silty, carbonate nodules (Fletcher, 1990). Several boreholes have been drilled in the Cow Ark area, both by the British Geological Survey as part of its mineral reconnaissance programme in 1975–1977 and, subsequently, by BP Minerals in 1981–1983. Two of the BGS boreholes, SD64NE/6 and 7, penetrated a few metres of grey to dark grey, sporadically calcareous and silty mudstone assigned to the (undivided) Bowland Shales; no marker bands were recognised. Some details are given by Wadge et al. (1983). Of the ten BP Minerals boreholes that are sited in the area, two (SD 64 NE/17 and 20) proved strata assigned to the lowest part of the Upper Bowland Shales, including the basal C. leion Marine Band. No indication has been obtained of the total thickness of the formation in this area. However, around Ashnott, nearby in the adjacent Clitheroe district, the thickness is about 90 m (Fletcher, 1990).

Grimsargh and Longridge Fell

In the south-eastern part of the district, the Upper Bowland Shales form a narrow outcrop some 13 km long, mostly along the face of the Longridge Fell escarpment. The heavily drift-covered part of the outcrop between Grimsargh and the northern part of Longridge town is conjectural, being neither exposed nor proved by boreholes. Farther north-east, discontinuous stream sections between Longridge and Ramsclough indicate that the total thickness increases towards the north-east from about 67 m in a section [6228 3926 to 6218 3934] (Bridge, 1989) south-east of Whitefold to about 115 m south-east of Brook House [6464 4082 to 6467 4070] (Aitkenhead, 1990a). The difference may be due in part to the necessarily arbitrary placing of the upper boundary in the passage into the overlying Pendle Grit Formation. A more accurate comparison can be made for the part of the sequence lying between the base of the C. leion and the top of the C. malhamense marine bands. This succession shows thicknesses of 62.77 m and about 73 m respectively in the two above-mentioned sections, which represents a much less-marked increase in thickness. Full details of the two sections are given by Bridge (1989) and Aitkenhead (1990a) in the appendices of their reports.

The basal C. leion Marine Band is best seen in the White Fold section where it consists of 5.13 m of tough, dark grey, calcareous mudstone with a petroliferous smell and a poorly preserved fauna of bivalves and ammonoids, including *Cravenoceras* sp. Another ammonoid fauna, including poorly preserved striatoid and girtyoceratid fragments, occurs in 2.2 m of platy mudstones 8.13 m above the C. leion Marine Band here; it probably also belongs to the E1a Chronozone. The E. pseudobilingue Marine Band is exposed in both the sections referred to above and in a gully [6236 3939] behind Bradley's Farm. In the Brook House section, where it is 3.5 m thick, the marine band is marked by conspicuous beds of grey, finely granular, argillaceous limestone and calcareous siltstone in a shaly scar on the hillside flanking the stream [6468 4080; 6461 4080]. In the White Fold section, 7.0 m of tough, platy, calcareous mudstones, including several bands with *Tumulites* sp., occur at this stratigraphical level. This section also provides continuous exposure of the interbedded dark fissile and tough platy mudstones between the E. pseudobilingue and C. malhamense marine bands, here totalling 13.10 m in thickness.

The C. malhamense Marine Band is also best exposed in the White Fold section, where it is 4.00 m thick. There are several other exposures in the area, some little more than scrapings made by

animals, revealing fragments of the characteristic brown decalcified fossiliferous mudstones. The easternmost exposure is in a cutting by a path [6640 4138] near Rakefoot and is calculated to lie about 20 m below the mapped base of the Pendle Grit (Fletcher, 1987). The White Fold section also shows the gradational upward change in the sequence, here only 4.08 m thick, between the C. malhamense Marine Band and the base of the Pendle Grit, a minimum for the Garstang district. The thinly interbedded, platy mudstones give way to paler fissile mudstones with a few beds of sideritic ironstone, up to 0.03 m thick, below the lowest sharp-based ferruginous sandstone bed assigned to the overlying formation.

MILLSTONE GRIT GROUP

Pendle Grit Formation

The Pendle Grit is much the most extensive Carboniferous formation at outcrop in the district. Of the three main outcrop areas the largest, with an area of about 60 km², forms the greater part of the Bowland Fells with extensions westwards to Nicky Nook and south-westwards to beyond Claughton. The much smaller outcrop (c. 4 km²) around Birkett Fell in the extreme east of the district mainly occupies the graben there. The third outcrop, some 13 km² in extent, forms Longridge Fell and probably continues westwards to Grimsargh under thick drift cover. Additionally, there is a small but conspicuous outcrop capping Beacon Fell in the Vale of Chipping. The best sections are shown in graphic form in Figure 11.

The name 'Pendle Grit' was first used by R H Tiddeman on the Geological Survey One-inch Old Series Sheet 92SE, published in 1878, for the sandstone (grit) capping Pendle Hill. It subsequently appeared on the first one-inch sheet of the present district, published in 1883. The name Pendle Grit Formation was introduced by Arthurton et al. (1988, p. 76) in the Settle district. The formation equates precisely with the Pendle Grit of Earp et al. (1961) in the adjacent Clitheroe district. Moseley (1962), referring to the sandstones overlying the Bowland Shales in the Sykes Anticline, used the term Pendle Grit on grounds of priority to replace his previously used (1954) term "Pendle Top Grit".

In view of the original usage, it is here proposed that the type section for the formation be designated as that exposed in Little Mearley Clough [7851 4110 to 7890 4123] on the north-west face of Pendle Hill in the Clitheroe district. Here, Earp estimated that nearly 500 feet (152.4 m) of the basal part of the formation are visible, together with the contact with the underlying Upper Bowland Shales (Earp et al., 1961, p. 118). The uppermost 46 m of the formation and the contact with the overlying Warley Wise Grit Formation are exposed in the gorge of the River Ribble near Salesbury Hall [6790 3631 to 6775 3595] (Bridge, 1988a). Other reference sections are mentioned below. In a recent basinwide study, Sims (1988) has recognised the same lithostratigraphic unit, with the same boundaries, but refers to it as the Pendle Grit Group.

In the Garstang district the Pendle Grit Formation is conformable on the Upper Bowland Shales and there is commonly a gradational coarsening-upwards change from the lower formation. In well-exposed sections the base is normally taken at the point where sandstone becomes predominant, i.e. forms over 50 per cent of the sequence.

The bulk of the formation in this district consists of medium- to coarse-grained feldspathic sandstone with subordinate interbedded silty mudstone and siltstone. The sandstone is usually in thick (0.3 to 4.5 m, average 1.5 m) irregular beds or, rather less commonly, in thinner (0.1 to 1.5 m, average 0.5 m) parallel beds. Beds in both facies are generally massive (i.e. internally structureless) with normal vertical grading either absent or weakly developed. Convolute, planar parallel and current ripple lamination structures are present towards the top of a few beds. Bases of beds are sharp, except where the base of a bed has amalgamated with the underlying one, a common occurrence in the irregularly bedded facies. Partings and thin intercalations of silty or argillaceous material are generally uncommon but, where they occur, sole structures are present on the bases of the overlying sandstone beds. Carbonaceous plant fragments and mudstone rip-up clasts are common in both facies. The two facies are not mutually exclusive but interdigitate with one another. In the Bowland Fells, the area where facies types can be most readily assessed, the irregularly bedded facies predominates and, in places, can be seen to truncate other facies erosionally.

Interbedded silty mudstone and siltstone form a significant proportion of the basal part of the sequence and in relatively thin, localised units widely and sparsely scattered through the formation. These minor units are soft relative to the adjacent sandstones and their presence is detectable by the negative topographical features ('slacks') they form in some places.

Sandstone in the argillaceous slacks and in the lowest part of the formation occurs in thinner (0.03 to 0.6 m, average 0.25 m) parallel beds and is finer grained, with a more argillaceous matrix than the two main facies. Howard (in preparation) notes that sole structures, including groove, flute and load casts, are fairly common. Interbedded mudstones may be either silty and fissile with laminae or very thin beds of silt, or fine sand ('striped beds'), or a poorly graded mixture of mud, silt and very fine to coarse sand with abundant dispersed mica and carbonaceous fragments.

Four representative samples (E61943, E61944, E61945 and E61946) of sandstone from the formation have been analysed and shown to be subarkoses with up to 17 per cent feldspar. K-feldspars and both untwinned and twinned plagioclase feldspars are recorded, with the first-named generally being most abundant (Appendix 2, Table 3). The plagioclase grains commonly show considerable dissolution along cleavages in sample E61946. Electron microprobe analyses on this sample show that the plagioclase is a pure albite (Ab_{97-100}) phase. Such pure compositions are typical of authigenic feldspars, of feldspars that have recrystallised during deep burial diagenesis as a result of the process of albitization, and of many feldspars that have been derived from low-grade metamorphic sequences and hydrothermally altered igneous rocks. Although albite overgrowths have been observed growing on detrital K-feldspar grains in the Pendle Grit by Sims (1988), widespread in-situ albitisation in the Pendle Grit is precluded by the co-occurrence of essentially unaltered K-feldspars and albite, the lack of albitisation (recrystallisation) textures such as blocky, discontinuous twin laminae or chessboard twinning, and the evidence for

dissolution of plagioclase grains (authigenic albite is stable during diagenesis). A detrital origin is therefore preferred for the untwinned albites in the Pendle Grit. They may have been sourced, at least in part, from a hinterland where low-temperature metagreywackes were exposed, such as the terrain to the north, where Ordovician and Silurian greywackes contain both albite and K-feldspar.

Detrital mica flakes (muscovite or biotite) form between 2 per cent and 4.6 per cent of the total mineralogy. The content of rock fragments, mostly mudstone or metapelite clasts, is low, invariably less than 2 per cent. However, detrital clay (chlorite or illite) is important, particularly in samples E61943 and E61944 from the basal part of the Pendle Grit, where relatively high amounts of clay matrix (up to 8.6 per cent) are present. Cements comprise quartz overgrowths, clays (kaolinite and, to a lesser extent, illite) and small amounts of iron oxides (hematite or goethite).

These findings are broadly in accord with the petrological account in the report on a basinwide study of the Pendle Grit by Sims (1988). This author's work contains no record of the presence of albite as a significant detrital constituent of the sandstones, however.

Fossils, except for carbonised plant fragments, are generally rare in the Pendle Grit Formation and in this district are restricted to the trace fossils *Gyrophyllites*, *Helminthoidea*, *Planolites* and *Protopalaedictyon*, together with other indeterminate forms (Howard, in preparation).

Although locally mappable sequences, mainly of silty mudstone, and sandstone with mudstone, have been delineated in places, the only formal subdivision of the formation is a local member, the Copster Green Sandstone (Bridge, 1988a). This lies in the middle of the Pendle Grit in the extreme south-east corner of the district, where part of its outcrop underlies the village of Copster Green [675 340]. The sandstone, which is medium to coarse grained, is distinguished by the presence of quartz pebbles and by the linear topographic feature which it forms within the Pendle Grit outcrop. Poor exposure precludes precise facies description; the mode of deposition of the member and the relationship to the parent formation are therefore problematical.

The sedimentological characteristics of the various Pendle Grit facies described above indicate deposition mainly from turbidity currents in relatively deep water. The thick, irregular-bedded facies resembles the coarse-grained turbidites of Lowe (1982), and was deposited by high-density turbidity currents. The truncating basal contacts observed in some sequences of this facies suggest that the turbidity currents were confined to channels, perhaps on a delta slope fan complex. The parallel-bedded facies, especially where there is a substantial interbedded argillaceous component, may have been deposited either in broader, shallower and less confined channels or in the interchannel parts of fans. The whole sequence probably represents a series of superimposed prograding submarine fans. The sparse palaeocurrent data collected during the present survey indicates a general current flow from the north.

These conclusions are in accord with a much more detailed and comprehensive basinwide study of late Pendleian (E1c) sedimentation by Sims (1988). This author's sedimentation model envisages deposition from turbidity currents in relatively long and narrow submarine channels, which are recognised mainly by their basal erosional contacts. They collectively formed a fan system prograding in a general southerly direction and accreting vertically rather than laterally, especially in the early stages of sedimentation when they were thought to have been constrained by relict 'high' areas within the basin. The channels terminated in small lobes at their distal ends, beyond which there was a short transition to basinal muds. The modern analogy for this model is cited as an active sand-rich fan in Bute Inlet, Canada (Prior et al., 1987).

The thickness of the formation, estimated from outcrop geometries, ranges from about 340 m in the west between Claughton and Calder Vale, to about 475 m on Longridge Fell (Bridge, 1988a). There are no borehole provings of the complete sequence, but a well at Oakenclough (SD 54 NW/1) proved the upper 208.03 m of the formation (see below).

Details

Bowland Fells, Nicky Nook and Claughton areas

The Bowland Fells area, being largely free of glacial deposits, provides numerous exposures in the extensive outcrop of the Pendle Grit Formation. Many of the stream sections are remote from public roads and difficult of access, however.

The gradational character of the base of the formation is well exposed in gullies on the main escarpment; examples are north-east of Hazlehurst [5742 4774], the west face of Parlick [5937 4534] (Howard, in preparation) and on Fair Oak Fell [6290 4740] (Howard, in preparation). Another section is exposed on the slopes above the right bank of Fiendsdale Water [5980 4938] in the heart of the Bowland Fells, where there is a tiny inlier of Upper Bowland Shales. Details are given by Howard (in preparation). Typically, there are a few laminae or thin beds of fine-grained sandstone in the uppermost 3 m or so of the Upper Bowland Shales, the proportion increasing upwards. The base of the Pendle Grit Formation is taken at the point where the proportion of sandstone appears to exceed 50 per cent. In this area, an appreciable argillaceous content (25 to 50 per cent) persists in this basal part of the formation which is delineated in places on the maps, though not given a formal name.

In the section north-east of Hazlehurst, mentioned above, some 82.5 m of this lower 'sandstone-with-mudstone' part of the formation are exposed. The following description is that of Howard (in preparation). The lower 21.5 m of the sequence consists of approximately 50 per cent sandstone and 50 per cent silty mudstone. The sandstone is generally feldspathic, micaceous and silty, with abundant carbonaceous fragments. It is also medium to coarse grained and poorly sorted. Beds average 0.20 m in thickness and are sharp based, with common sole structures, particularly flute and load casts; most are internally massive, some showing normal vertical grading with parallel laminations. The interbedded mudstone is either fissile, with thin layers of siltstone, or massive, very silty and micaceous, with abundant carbonaceous plant fragments. The upper 61 m comprises about 75 per cent sandstone and 25 per cent mudstone. The sandstone is feldspathic and slightly micaceous, with a grain size ranging from fine to very coarse. Sedimentary structures include normal vertical grading and parallel, convolute, ripple and climbing lamination. Individual beds average 0.40 m in thickness and have sharp bases with groove, flute and load casts. At many levels the sandstone beds are amalgamated, with mudstone flakes common along amalgamation surfaces. The interbedded silty mudstone is typically fissile, with common carbonaceous plant fragments and thin layers of siltstone or very fine-grained sandstone. Another excellent exposure of this lower sandstone-with-

mudstone sequence occurs in the deep gully on Fair Oak Fell, mentioned above and described in detail by Howard (in preparation).

Some 91.5 m of the basal part of the main Pendle Grit sequence is exposed in a tributary of Fiendsdale Water [5992 4923 to 5993 4913], west of Bleadale Ridge, where about 67.1 m of the underlying sandstone-with-mudstone facies is also exposed. The coarse- to very coarse-grained, irregularly bedded facies predominates in this section, but in the top 5.0 m an intercalated sandstone-with-mudstone sequence occurs. This sequence is possibly the same as that present over the watershed to the south-west, in a stream section on Holme House Fell [5880 4798 to 5887 4814], which additionally includes 25.15 m of overlying strata consisting mainly of fine- to very coarse-grained, irregularly bedded sandstone. Elsewhere in the Bowland Fells there are numerous exposures of these sandstones in streams and gullies, and in crags forming the scars of some of the landslips in the area, such as Scout Rock [6224 4967] and Holdron Castle [6100 5078]. Some details are given by Howard (in preparation) and Hughes (1986, 1987).

In the western part of the area the River Calder has cut a gorge through the till-covered Pendle Grit outcrop and there are several good exposures in meander scars between Sandholme Mill and Calder Vale [e.g. 5188 4350; 5245 4361 to 5210 4260; 5281 4393; 5338 4600]. These are too scattered for a continuous sequence to be built up, however. A water borehole (SD 54 NW/1) at Oakenclough started near the top of the formation and proved a sequence consisting predominantly of medium- to coarse-grained sandstone with many thin bands of mudflakes and a few interbedded argillaceous beds up to 6.1 m thick, to the bottom of the hole at 213.36 m.

The highest beds in the sequence are exposed in a strike section on the left bank of the Grizedale Brook [5175 4808], but the top of the formation is not seen. The section at this locality is 16.8 m thick and consists of massive and laminated, sharp-based sandstone beds with shaly mudstone intercalations, and includes a 2.7 m slumped unit 8.6 m above the base. The overlying Cravenoceras cowlingense Marine Band is exposed a little further downstream (see p.53). To the north, numerous small exposures of mostly massive, medium- to coarse-grained sandstone are located on the upstanding promontory formed by the Nicky Nook Anticline.

A little beyond the farthest downstream exposure [5181 4343] near Sandholme Mill, the outcrop of the Pendle Grit is displaced by the Grimsargh Fault. To the south-west of this fault the outcrop lies under thick drift and is largely conjectural, the only indication of its existence being in a borehole (SD 54 SW/23) near Ducketts Farm. This proved 15.85 m of drift overlying a sequence of fine- to coarse-grained sandstone interbedded with mudstone and siltstone (approximately 70 per cent sandstone and 30 per cent mudstone). The sandstone is poorly bedded and mainly massive, with some laminated beds and bands of mudflakes. Proximity to the basal Permo-Triassic unconformity is indicated by a purplish tinge.

Birkett Fell

In the north-eastern part of the district the Pendle Grit occupies an area of high moorland in the Birkett Fell graben between the Thorneyholme, Giddy Bridge and Browsholme Moor faults (Figure 21; Fletcher, 1990). Its thickness is probably about 110 to 150 m, and it is presumed to represent the lower part of the formation. In the Waddington Fells area in the adjacent district to the east, however, the Pendle Grit is either greatly attenuated over what was a 'high' in the basin floor (Sims, 1988) or has been largely removed by erosion prior to the deposition of the Warley Wise Grit Formation (Poole in Earp et al., 1961). This problem has not been reassessed during the present survey and remains unresolved. The predominantly sandstone sequence in this area differs noticeably in two respects from that in the main part of the Bowland Fells. Firstly, about 25 m of very coarse-grained, massive, pebbly sandstone is present in the uppermost part of the outcrop. This occurs mainly in a belt, about 250 m wide, aligned parallel to the Browsholme Moor Fault and extending from around an exposure [6794 4953] near Knowlmere Manor to another exposure [6839 4796] north-east of Crimpton. Secondly, the sequence was intensely reddened due to its former proximity to the basal Permo-Triassic unconformity, and then irregularly leached by acid waters percolating down from the ground surface in Flandrian times.

Lithologies in the main part of the sequence are largely of the parallel-bedded turbidite facies, with subordinate thick, irregular beds and fairly common mudstone and siltstone interbeds. The sequence is best exposed in Birkett Brook [6779 4633 to 6798 4839] and one of its tributaries [6739 4880 to 6801 4892]. Interbedded units of red and red-grey micaceous silty mudstone form slacks in several places, notably on the north-west side of Birkett Fell [669 481]. The thickest unit exposed, some 7.0 m thick, occurs in the stream [6784 4884] draining Hodder Bank Fell. The very coarse-grained, pebbly facies at the top of the sequence is best seen in the lower reaches of Birkett Brook [6823 4865 to 6827 4908]. At the former locality, the massive pebbly sandstone forms a 13 m-high scar. The feldspars are highly decomposed and the sandstone has a granite-like appearance, due to its coarse massive character and the pale pink to white colouration produced by deep weathering and leaching (Fletcher, 1990). Similar sandstone has been quarried on Kitcham's Hill [6700 4822] on the highest part of Birkett Fell. Fletcher (1990) has pointed out that similar sandstones exposed on Marl Hill Moor [6888 4693] in the adjacent Clitheroe district are cross-bedded, a facies assigned by Poole (in Earp et al., 1961) and Sims (1988) to the overlying Warley Wise Grit Formation.

Longridge Fell and Ribblesdale

In the south-east, the Pendle Grit crops out around the Ribchester Syncline. Only the north-west limb and part of the south-east limb lie within the present district, however. The estimated thickness here is about 475 m (Bridge, 1988a), a maximum for the district and probably near the maximum for the basin of deposition as a whole (Sims, 1988). The south-western part of the outcrop lies beneath thick drift between the Grimsargh and Longridge faults; its subcrop is conjectural and unproven by boreholes. East of the latter fault, the prominent feature formed by the Pendle Grit outcrop rises abruptly within the eastern part of Longridge town from about 110 to 140 m above OD.

The base and lowest 4.70 m of the formation are exposed in a stream gully [6218 3925] south-east of White Fold (Bridge, 1989). Much of the basal 73.5 m is in the parallel-bedded sandstone-with-mudstone facies and this gives rise to a persistent slack on the hillside below the main sandstone escarpment between Longridge and Cardwell House. Several other slacks, probably representing similar more argillaceous sequences higher in the formation, have been delineated both on the scarp and dip slope, i.e. both on the north-west and south-east slopes of Longridge Fell. The highest slack, between the top of the Pendle Grit and the base of the Warley Wise Grit, is present in the eastern part of the outcrop but thins westwards and is not mappable beyond Hoardsall [637 387]. The beds forming the slack are best exposed in the River Ribble section near Salesbury Hall where they are about 63 m thick (Figure 11, column 8). This section is given in detail by Bridge (1988a, pp.23–25).

The sandstones that predominate in the middle part of the formation are well displayed in disused quarries just east of Longridge, where a composite discontinuous section has been measured in detail by Bridge (1989, pp.23–26). The quarries include Lord's Quarry [6126 3809], Spencer's Quarry [6142 3807], Tootle Height [6143 3796] (Plate 10) and Copy [6195 3835]. The section (Figure 11, column 6, top) has a total thickness of 173.45 m, of which about two thirds is exposed. The sandstones display both parallel and

amalgamated bed forms and are mainly medium grained. A few coarser beds show normal grading. Towards the top of the sequence in the Tootle Height Quarry a few irregular beds of sandstone contain quartz pebbles and mudstone flakes, suggesting a possible correlation with the Copster Green Sandstone (see below). Several of the thicker beds of sandstone in Spencer's Quarry and Lord's Quarry in the middle part of the section contain ovoid, concretionary, ferruginous, carbonate-cemented bodies, or 'doggers', rarely found elsewhere in the district.

Farther east, a disused quarry [6365 4013] near Cardwell House provides a good 9 m section of parallel bedded, medium-grained sandstone not far above the base of the formation (Aitkenhead, 1990a). Three quarries [6755 4065; 6785 4065; 6825 4058] (Plate 11) north of Fell Side Farm near the eastern margin of the district lie along the strike of planar, thickly bedded, medium-grained sandstones in the middle part of the formation. Another group of small disused quarries situated farther down dip to the south, and therefore somewhat higher in the succession, display coarser-grained sandstone with irregular and amalgamated beds, including some with scattered quartz pebbles and mudstone clasts. Examples include Intack Quarry [6627 3954] and a quarry [6821 3977] north-east of Higher Deer House.

In the south-east corner of the district, the gorge (Sales Wheel) of the River Ribble near Salesbury Hall [6790 3631 to 6756 3586] provides a good section of both the Pendle Grit and the overlying Warley Wise Grit (Bridge, 1988a). The highest part of the formation here [6780 3599 to 6776 3595] consists of a 54.55 m sequence of interbedded silty mudstone, siltstone and sandstone which is mappable for several kilometres to the north, but cannot be delineated far to the south owing to the thick drift cover. Upstream [6790 3630 to 6780 3590] some 19.1 m of the underlying beds are exposed and are included in the undivided, predominantly sandstone part of the formation. The sandstone is fine- to medium-grained and occurs mostly in laminae or thin beds intercalated with micaceous silty mudstone and siltstone.

The Copster Green Sandstone is estimated by Bridge (1988a) to have a thickness of about 135 m in the Ribble Gorge, reducing to about 100 m farther south around Clayton-le-Dale. It forms a well-defined but discontinuous feature from just south of the Ribble [682 363] to Clayton-le-Dale. The best exposures are in small disused quarries at Clayton-le-Dale [6719 3321] and Copster Hall [6767 3421]. The sandstones are medium to coarse grained, sparsely pebbly to pebbly and lack clearly defined internal lamination, although traces of wavy bedding and cross-lamination have been noted (Bridge, 1988a).

Warley Wise Grit Formation

In the south-eastern part of the district, the Pendle Grit Formation is overlain by another major sandstone-dominated unit, the Warley Wise Grit Formation, first described by Bray (1927) from around Warley Wise, the type area in the Clitheroe district (Earp et al., 1961, p.105). It now seems likely that the Warley Wise Grit and the Brennand Grit of the districts to the north belong to essentially the same delta top sand body of late Pendleian (E1c) age as the Grassington Grit Formation, first formally named by Dunham and Wilson (1985, p.62) in the type area of the Askrigg Block. The correlation of the Wilpshire Grit, as mapped in the adjacent Preston district to the south, is slightly less certain, but it also is probably coeval with the Grassington Grit. The Lower Wilpshire Grit of the Preston district is almost certainly coeval with the main Pendle Grit Formation sandstone.

Plate 10   Sandstone of the Pendle
Grit Formation exposed in a 20 m
face in the disused quarry at Tootle
Height [6143 3796], Longridge. The
sandstones are medium to coarse
grained with quartz pebbles and
mudstone clasts in some beds, and
there are some impersistent dark,
shaly mudstone intercalations. The
beds mostly appear internally
structureless and have sharp erosive
bases (A14816).

In the present district, Riley (1985) has identified the basal
Arnsbergian (E2a1) Cravenoceras cowlingense Marine Band
immediately overlying the Warley Wise Grit in the River
Ribble section (Sales Wheel) near Salesbury Hall [6775
3595], proving the late Pendleian age of this formation.
About threequarters of the full thickness of the formation,
measured as 117.18 m by Bridge (1988a), is exposed here,
and the section is therefore proposed as a local reference
section.

The outcrop is not easy to distinguish from that of the
Pendle Grit and it becomes increasingly conjectural to the
west of Knowle Green [637 381] and south-west of White
Holme [668 337] as features become more and more subdued, presumably due to the thickening drift cover.

The one good section of the Warley Wise Grit near
Salesbury Hall (see below) is of a facies that includes both

mass-flow and cross-bedded elements. Following a reconnaissance study of the whole Grassington Grit and Warley
Wise Grit outcrop, Sims (1988) concluded that these formations represent coarse sediment deposited in a south to south-
westward prograding braid-delta in which a basal deeper
water mass-flow apron was overlain by large-scale cross-
bedded sands and then by fluvial deposits.

Details

River Ribble and area to the north

A detailed measured section of the Warley Wise Grit in the gorge of
the River Ribble [6775 3595 to 6756 3586] near Salesbury Hall, is
given by Bridge (1988a). Three distinct and locally mappable leaves
are recognised here, each consisting of coarse-grained sandstone
with some pebbly beds, separated by sequences mainly of thinly

Plate 11 A working quarry (Leeming Quarry) in thickly bedded, medium-grained sandstones of the Pendle Grit situated [6825 4058] on a long dip slope on the south-east flank of Longridge Fell (A14771).

interbedded or interlaminated siltstone and sandstone. These sub-divisions cannot be traced in the thickly drift-covered country north and south of the river, but north of Bailey Hall [678 373] they become recognisable again by their associated topographical features.

In the river section, only the upper sandstone division shows clear, well-developed cross-bedding (Plate 12). The sandstone beds in the middle leaf are either massive (without internal lamination) or show weakly developed, low-angle cross-stratification, while those in the lowest leaf appear massive.

Some of the best exposures are in the banks of the Dean Brook [6829 3746 to 6812 3862]. A disused quarry at the last locality exposes 14.5 m of massive, medium- to coarse-grained, poorly cemented sandstone with dispersed pebbles, that readily dis-aggregates to sand. Exposures farther west, as on Doe Hill [6771 3821] and north-east of Huntingdon Hall [6637 3917], are mostly small diggings in cross-bedded pebbly sandstone, but beyond here the gorge of the Duddel Brook provides better sections. The best of these is in a cliff on the west bank [6528 3897] which exposes 7.5 m of medium- to coarse-grained sandstone with scattered pebbles in beds up to 1.2 m thick.

Sabden Shale Formation

The Sabden Shales (Bisat, 1924) occur only in the south-east of the district, in the Ribchester Syncline. Strata of the same age in the northern outcrop contain several thick sandstones, and the name Sabden Shales does not apply (Figure 10). However, this unit is fully present in the Preston district to the south, where it lies between the top of Warley Wise (Wilpshire) Grit and the base of the Parsonage Sandstone, and spans the long interval between the basal Arnsbergian and Middle Kinderscoutian stages (Price et al., 1963). These authors imply (op. cit., fig. 3) a group status for the Sabden Shales, but it is now formally proposed as a formation, with the section at Salmesbury Bottoms, described in detail by Moore (1930b), designated as the type section. The informal subdivision into Lower, Middle and Upper Sabden Shales has little practical value and has therefore been abandoned.

A total thickness in the order of 400 m is estimated to be present in the Ribchester Syncline, ranging in age from basal Arnsbergian (E2a1) to Chokierian (H1) and possibly

Plate 12   Warley Wise Grit: trough
cross-laminated, coarse-grained
pebbly sandstone near the top of the
formation, exposed on the left bank
of the River Ribble [6757 3584] near
Salesbury Hall (A14770).

Alportian (H2), but the top is not present. There are no
borehole provings of any stratigraphical value.

As the name suggests the formation is predominantly
argillaceous but, locally, north-east of Ribchester, it contains
a high proportion of siltstone and sandstone. Most of the succession is exposed in three major sections, but these give no
indication of lateral variation because no one section
overlaps with another. The basal 50 m consists of dark grey
mudstone with bands containing marine fossils, alternating
with slightly paler barren mudstones and silty mudstones
with scattered thin beds and small nodules of sideritic
ironstone. There are four important marine bands, those of
Cravenoceras cowlingense (E2a1), *Eumorphoceras ferrimontanum*
(E2a2) (formerly known as *E. bisulcatum*), *E. yatesae* (E2a3)
and *Cravenoceratoides edalensis* (E2b1). A fifth band, that of *Ct.
nitidus* (E2b2) may be present at the top of this basal sequence but has not yet been confirmed (see below). The
middle part of the sequence, inferred to overlie these mudstones, includes 209 m of siltstones and interlaminated or
thinly interbedded fine-grained sandstones with several
erosively based coarser sandstone packages, three of which
are locally delineated on the maps. Mudstones with marine
faunas overlain by turbidite sandstones and siltstones occur
in the highest exposed part of the succession. The marine
bands include that of *Nuculoceras nuculum* (E2c4), the highest
in the E2c Chronozone, and two bands with *Isohomoceras
subglobosum* in the lower part of the succeeding H1a
Chronozone.

The marine basin in which the Sabden Shales were
deposited was apparently subjected to an influx of sand and
silt during the late Arnsbergian interval (E2b2 – E2c4); there
are no marine bands to provide biostratigraphical control.

The interpretation of the arenaceous beds is somewhat problematical because of the lack of positive evidence, such as grading or cross-bedding, that would indicate the nature of the depositing currents. On balance, the general monotony of the sequence and lack of clear fluviatile indicators leads to the tentative conclusion that the sediments were deposited in a delta slope environment.

Details

Most of the Sabden Shales succession in this district is exposed in three widely separated sections: in the River Ribble near Salesbury Hall, in the gorge of the Duddel Brook, and in the River Ribble near Balderstone Hall. The main lithologies and faunas present are given in Figure 12. The Salesbury Hall ('Sales Wheel') section [6756 3586 to 6746 3585] has long been noted for its dark grey fossiliferous mudstones overlying the thick sandstone sequences of the Warley Wise and Pendle Grit formations (Phillips, 1836, p.235; Moore, 1936, p.190). The following description is based on that of Riley (1985) who measured and collected it at a time of exceptionally low river level.

The dark grey mudstones in the basal 8.7 m are sulphurous and fissile with a few thin sideritic ironstone bands and nodules. Though the eponymous goniatite has not been found, the faunal assemblage of *Posidonia corrugata* and conodonts, found just above the base, is good enough for the horizon of the C. cowlingense Marine Band (E2a) to be inferred in the context of the section as a whole. These beds are succeeded by fossiliferous, brown-weathering, calcareous siltstone with fissile mudstone intercalations, forming the E. ferrimontanum Marine Band (E2a2). This band has been renamed since Riley's (1985) account in order to avoid confusion with other Arnsbergian marine bands containing *Eumorphoceras* species formerly referred to *E. bisulcatum* (Riley, 1987). The E2a2 band is overlain by 33 m of mainly fissile mudstones with ironstone bands, some incompetently folded, and these in turn by the E. yatesae Marine Band (E2a3). The marine faunas in this band, including *E. yatesae*, are distributed sparsely through 2.2 m of strata, which include a thin ironstone band near the base. Sideritic ironstone bands and nodules are also present in the overlying 7 m of mudstone up to the base of the *Cravenoceratoides edalensis* Marine Band, which is 1.75 m thick near the top of the section. This band includes large limestone concretions (bullions) in the basal part.

The Duddel Brook section encompasses some 208 m of siltstones with three mappable sandstone units. The latter are massive, coarse and pebbly in the lower part with a discordant erosive base, suggesting rapid deposition from suspension in a channel. The finer-grained sandstones and siltstones provide little positive evidence of their depositional environment and their sedimentology has not been studied in detail. The correlation of this section with the general succession is also somewhat problematical since no faunal marker bands have been found. Similar beds with a total thickness of about 123 m are exposed in the Stydd Brook to the west [6508 3706 to 6518 3672], where the section is complicated by strike faulting (Bridge, 1988a).

On the basis of general field relations, the sandstones exposed in the Duddel and Stydd Brook sections are assumed to lie in the E2b1 – E2c3 part of the Sabden Shales sequence (Figure 12). There are no faunal marker bands.

The highest beds in the local Sabden Shales sequence are exposed in a discontinuous section in the River Ribble near Balderstone Hall, partly on the left bank [6131 3330 to 6147 3301] and partly on the right bank [6153 3305 to 6157 3300] (Figure 12). They include the topmost Arnsbergian (E2c4) marine band containing *Nuculoceras nuculum* and the overlying basal Chokierian (H1a1) band with *Isohomoceras subglobosum*. *N. nuculum* has also been found upstream at

Old Park Wood [6522 3435] and in a nearby tributary stream [6516 3403]. It is not certain which of the three *N. nuculum* levels normally present in the complete E2c sequence are represented here, however. The Balderstone Hall section also exposes the two lowest Chokierian bands with *Isohomoceras subglobosum* (H1a1 and H1a2). Again, other isolated exposures yielding this goniatite in Old Park Wood [6520 3418; 6505 3429] cannot be correlated precisely with any one of the three levels where this species usually occurs.

Higher strata in the Balderstone Hall section include sandstones with features characteristic of turbidites, such as sharp erosive bases, flute casts, mudflake conglomerates and graded bedding. These sandstones together comprise a mappable member that can be traced for about 2 km north-east and south-west of Balderstone Hall.

Roeburndale Formation and Caton Shale Formation

In the north-west part of the Garstang district the Pendle Grit is generally succeeded by the Roeburndale Formation (Figure 10). The intervening Brennand Grit Formation (analogous in position to the Warley Wise Grit, see above) is only locally present near Hawthornthwaite Fell, where it appears to die out shortly after crossing the district boundary. The term Roeburndale Formation was introduced by Arthurton et al. (1988) to replace the 'Roeburndale Grit Group' of Moseley (1954). The formation boundaries chosen are those in Moseley's later (1956) paper, with the base at the top of the Brennand (Grassington) Grit and the top at the base of the Caton Shales. In the Brennand Fell area of the adjacent district to the north (Brandon, in preparation) the Brennand Grit is overlain by a mudstone sequence, the 'Tarnbrook Wyre Marine Beds' of Moseley (1954), in which the Cravenoceras cowlingense Marine Band (E2a1) lies about 70 m above the base. This sequence is included in the Roeburndale Formation, thus extending its age range down into the late Pendleian (E1c). This is older than the age previously published (Arthurton et al., 1988, p.81) though the bulk of the formation remains early Arnsbergian (E2a) in age. In the present district the Brennand Grit is absent and the argillaceous sequence containing the Cravenoceras cowlingense Marine Band immediately overlies the Pendle Grit. The topmost part of the formation in the Lancaster district type area is characterised by the presence of ganisteroid sandstones and a coal seam (Moseley, 1954, p.431). This part of the sequence is little exposed in the present district but, because the general character of the formation here is turbiditic rather than fluviatile, it is unlikely that sediments of subaerial facies are present. The total thickness of the formation is estimated to be about 450 m.

Several sandstone members have been mapped within the formation, both to the north and the south of the Nicky Nook Anticline. These comprise the **Park Wood Sandstones** which lie between the C. cowlingense (E2a1) and the E. ferrimontanum (E2a2) marine bands, the **Stonehead Sandstone** above the E2a2 band, and an unnamed sandstone at the top of the formation.

The Park Wood Sandstones Member is named after Park Wood, Wyresdale Park [507 493], where it forms a broad feature. It is poorly exposed but probably consists of several sequences of turbiditic sandstones interbedded with silt-

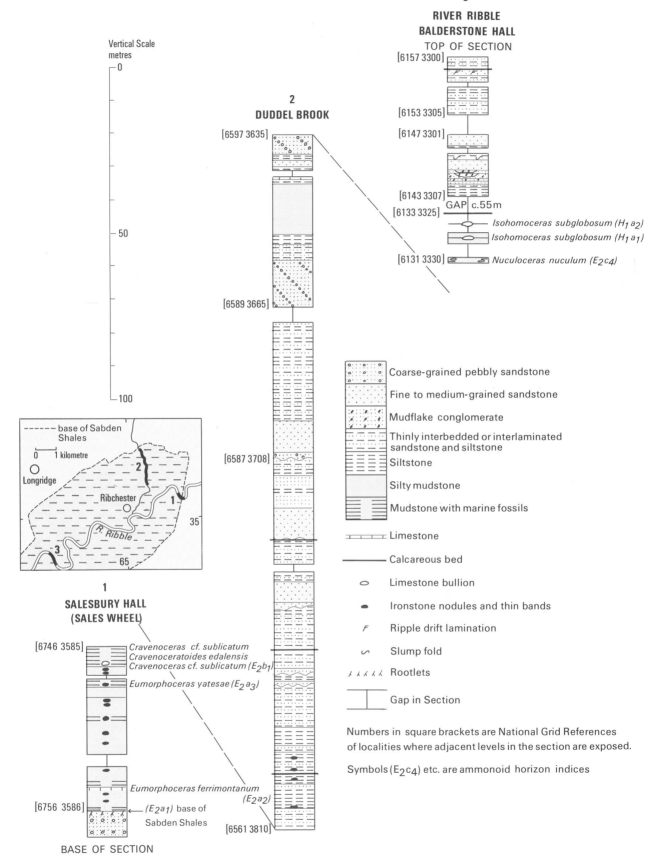

Figure 12   Sections of Sabden Shales in the Ribchester Syncline.

stones and silty mudstones. The stratotype section is in Grize Dale (see below). The total estimated thickness is about 250 m.

The Park Wood Sandstones are overlain by a predominantly mudstone sequence that includes the most widely identified and useful early Arnsbergian (E2a2) marker, the Eumorphoceras ferrimontanum Marine Band. In this area there is an unusual development of this band, consisting of dark grey calcareous mudstone, only about 0.7 m thick, intercalated between two turbiditic calcareous sandstone beds. The overlying Stonehead Sandstone is named after a farm at the northern boundary of the district [5300 5210]. This member is again of turbiditic facies and is up to 55 m thick. A thin sandstone unit is probably present at the top of the Roeburndale Formation but is very poorly exposed and the outcrop, being mainly under drift, is largely conjectural.

The Caton Shales, whose type area lies in the Lancaster district to the north, are estimated here to be about 50 to 70 m thick. There is only one exposure (see below) and this shows the dark grey, fissile fossiliferous mudstones characteristic of the formation (Moseley, 1954; Arthurton et al., 1988).

Details

The basal contact of the Roeburndale Formation on the Pendle Grit Formation is not exposed. The lowest beds seen are opposite Pedder's Wood in Grize Dale, on the steep left bank of the Grizedale Brook [5117 4782], where they are estimated to lie about 30 to 50 m above the base. The section includes the basal Arnsbergian C. cowlingense Marine Band and the base of the overlying Park Wood Sandstones; it is as follows:

	Thickness m
Sandstone, thinly to thickly bedded, mainly massive but with some parallel lamination, and intercalated mudstone, some reddening adjacent to joint planes	5.27
Mudstone and discontinuous sandstone beds, thinly interbedded; some slumping	6.00
Mudstone, blocky, with thin ironstone bands and nodules	8.00
Mudstone with thin discontinuous and slumped sandstone beds	4.60
Mudstone, dark grey fissile, fossiliferous, with a 0.23 m grey crystalline limestone at the base, truncated laterally by slumping. Marine fauna includes P. corrugata, Anthracoceras or dimorphoceratid fragments, Cravenoceras cowlingense and fish debris	0.90
Mudstone, dark grey	1.30
Mudstone, grey, with some sideritic ironstone and pyritic nodules and a few thin beds of sandstone	12.58

The Park Wood Sandstones Member forms a broad feature in Wyresdale Park where, in Park Wood, a degraded and overgrown quarry [5072 4940] exposes some 4.0 m of tough quartzose sandstone in thick sharp-based beds apparently lacking internal laminations. Petrographical data from one thin section (E61949) of this sandstone are given in Appendix 2, Tables 3 and 4. Similar but thinner-bedded and flaggy sandstones were recorded in the nearby M6 motorway cutting to the south-west. Site-investigation boreholes drilled hereabouts prior to motorway construction indicate a high proportion of interbedded mudstone and siltstone in parts of this thick but little known sequence. The member is probably present in two leaves on the west side of the Calder Valley near Oakenclough but cannot be traced very far to the south.

The mainly mudstone sequence, with the E. ferrimontanum Marine Band, between the Park Wood Sandstones and the Stonehead Sandstone forms a broad outcrop in the Barnacre Syncline around Barnacre Lodge [5146 4654], to the west of the Calder valley. The wide outcrop reflects the presence of a long gentle westerly inclined dip slope. The sequence is best seen in the streams south of Barnacre Lodge [5158 4638 to 5116 4603] which have eroded through the thin till cover into the beds around the E. ferrimontanum level. The marine band is exposed at both these referenced localities and consists of dark grey, platy, calcareous, fossiliferous mudstone, 0.69 to 0.73 m thick, sandwiched between two hard turbiditic sandstone beds (Plate 13). The fauna includes: P. corrugata, Cravenoceras sp., E. erinense, E. cf. ferrimontanum, Kazakhoceras scaliger and crinoid fragments.

The thickness of fossiliferous mudstone is less than normal, probably due to the erosive effects of the turbidity currents that deposited the two adjacent sandstone beds. Petrographical analyses of samples of the underlying (E61947) and overlying (E61948) sandstone beds, by B Humphreys, show that they are calcareous subarkoses with over 20 per cent Fe-carbonate cement. Feldspars constitute c.10 per cent of the rock with K-feldspars slightly in excess of plagioclase. Despite the pervasive carbonate cement, there is clear evidence for early diagenetic feldspar dissolution predating emplacement of the carbonate cement. X-ray diffraction proves that the cement is largely ferroan calcite with minor dolomite. Some fine-grained carbonate cement, probably siderite, also occurs.

Major, widely developed marine bands such as this are generally regarded as representing periods of marine transgression. Such transgressions can effectively drown delta floodplains and would be expected to restrict sand supply to turbidite fans on the prodelta slopes. The fact that such clastic input has locally continued during the deposition of this marine band implies either that tectonic effects in the source area have overidden the transgressive dampening or that the source area hinterland was not a wide, easily drowned floodplain but an area of much higher relief, with a steeper gradients.

The Stonehead Sandstone in the Barnacre area forms two main outcrops, east of New Hall Farm [511 457] and around Kelbrick Farm [530 466], south-west of Oakenclough. There are a number of exposures near New Hall Farm, notably by the road [5100 4592] immediately to the east and on the opposite side of the road in a disused quarry (Holker's Quarry) [5108 4606], both showing about 3 m of unlaminated, thickly bedded, medium-grained sandstone. Three hillocks on the long slope to the east and north-east are interpreted as outliers of the Stonehead Sandstone; the two southern outliers [522 458 and 525 458] are entirely drift covered. However, the northern one [5206 4734], east of Burns Farm, has a well-exposed section in a disused quarry. This consists of 10.3 m of pinkish grey, medium-grained sandstone in thick, unlaminated, irregular, amalgamated beds.

In a stream [5059 4962 to 5050 4965] in the north-west corner of the district, south-west of Park Gate, some 9 m of massive, thickly bedded sandstone are exposed, together with several sections in the overlying grey, silty, fissile mudstones which contain scattered bands of ironstone nodules. The exposure at the type locality [5323 5192] near Stonehead Farm at the northern boundary of the district consists of 4.5 m of thinly to thickly bedded, fine- to medium-grained sandstone in which mica flakes and carbonaceous fragments are conspicuous. There are also several other sections in the member, for example one in a stream south of Cliftons [5140 5004], that consists of thinly interbedded sandstone, siltstone and silty mudstone.

The unnamed sandstone assumed to lie at the top of the Roeburndale Formation is seen only in the banks of the small stream [5188 4753] north-west of Burns Farm, where 1.5 m of

Plate 13   The *Eumorphoceras ferrimontanum* Marine Band sandwiched between two turbidite
sandstone beds in the stream beside All Saints Church, Barnacre [5116 4604]. The dark,
calcareous, fossiliferous mudstone of the marine band, 0.69 m thick, is at the level of the hammer,
beneath the sharp erosive base of the overlying sandstone bed (A14795).

massive, medium-grained sandstone are exposed.

There is only one good exposure of the Caton Shales in the
district. This occurs in the left bank of the Grizedale Brook [5074
4726] to the north-east of Woodacre Hall and was measured by Dr
Riley as follows:

	Thickness m
Mudstone, fissile, weathered	4.00
Mudstone, fissile, with impersistent ironstone at top; fauna includes *P. corrugata* and *Anthracoceras* or dimorphoceratid ammonoids	1.00
Mudstone, fissile, weathered	2.00
Mudstone, fissile, decalcified, with *P. corrugata* and *Anthracoceras* or dimorphoceratid ammonoids	0.05
Mudstone, fissile, apparently barren	0.10
Mudstone, decalcified, with a shaly bullion 1.50 m above base; fauna includes *P. corrugata, Selenimyalina variabilis, Anthracoceras* or dimorphoceratid and cravenoceratid ammonoids	2.05
Clay, pale bluish grey (K-bentonite)	0.003

Mudstone, fissile to blocky, with marine fauna mainly 2.9 m above base including *S. variabilis erecta, C. holmesi* and *Cravenoceratoides nitidus*	3.90
Mudstone, black, fissile, barren	2.00

These typically fossiliferous mudstones are of E2b2 age and
probably lie in the middle part of the Caton Shales. K-bentonite
clay bands are known at various levels in mudstones of Arnsbergian
to Kinderscoutian age, particularly in the south Pennine region
(see, for example, Aitkenhead, 1977). They are thought to be
deposits of fine wind-borne volcanic tephra (Trewin, 1968).

Strata above the Caton Shales

An estimated 700 m or so of strata overlying the Caton
Shales (Figure 10) are inferred to crop out in the north-west
corner of the Carboniferous outcrop. They mostly lie
beneath thick drift and are largely unproved. Much of the sequence is exposed in the adjacent district to the north,
however, and is described in a report by Wilson et al. (1989).
In ascending order, the sequence comprises: the Heversham

House Sandstone Formation, the Crossdale Mudstone Formation, the Wellington Crag Sandstone, the Dolphinholme Mudstone and the Ellel Crag Sandstone Formation.

The **Heversham House Sandstone** may be up to 120 m thick. It appears to be of mixed facies in the type area to the north, with a seatearth sandstone at the top. The mainly turbiditic assignation shown in Figure 10 tentatively assumes a transition to deeper water facies towards the south-west. The **Crossdale Mudstone Formation** is about 160 m thick and, in the adjacent district, contains the widely occurring and stratigraphically important *Isohomoceras subglobosum* fauna (Wilson et al., 1989). The **Wellington Crag Sandstone** is about 35 to 60 m thick and, in the adjacent district, is coarse-grained and cross-bedded. The succeeding mudstone unit, the **Dolphinholme Mudstone**, probably includes beds of Chokierian to Alportian age, on the basis of exposures in the type area, which yield a poor bivalve fauna. The unit is estimated to be about 60 m thick. The **Ellel Crag Sandstone Formation** is about 60 to 90 m thick and is likewise known only from the type area in the adjacent district, where it is fine to medium grained, with large-scale cross-bedding. Some 240 m of unnamed strata, probably mainly of Kinderscoutian age but possibly younger, overlie the Ellel Crag Sandstone. The sequence includes one medium- to coarse-grained sandstone member, estimated to be about 35 m thick, which is exposed near Forton (see below).

Details

Exposures are rare in this largely drift-covered area of Wyresdale. The only boreholes are those drilled during site investigations for the M6 motorway, and these are of little stratigraphical value. The Wellington Crag Sandstone forms a broad feature north-west of the River Wyre. The top of this feature around Gleaves Hill [5080 5204] is free of drift and there are small exposures of cross-bedded, brown, coarse-grained sandstone.

Farther west, a disused quarry (Richmond Quarry) [5862 5205], north of Forton, exposes an unnamed sandstone member, the highest in the sequence. The exposure consists of 4.6 m of thickly bedded, medium- to coarse-grained sandstone with large-scale cross-bedding.   NA

FOUR

Permian and Triassic rocks

Permian and Triassic strata, which crop out in the western half of the district (Figure 1), are overlain by an extensive cover of drift. Numerous boreholes have been drilled in the Sherwood Sandstone for water abstraction and monitoring purposes, especially since the Second World War. In the Mercia Mudstones, on the other hand, boreholes are very sparse; the intensive drilling for salt which took place in the adjacent Blackpool district has not been matched here because of the south-eastwards thinning of the salt beds (Wilson and Evans, 1990, fig.6). The most recent phase of deep geological exploration has been associated with the search for hydrocarbons. This has been mainly by seismic reflection methods, but the first wildcat well was drilled for British Gas at Thistleton (SD 33 NE/17) early in 1988 and another is planned at Hesketh in the Preston district. The results remain confidential for commercial reasons but are likely to provide important additions to our geological knowledge.

CLASSIFICATION

Such evidence as there is indicates that the sequence below the Sherwood Sandstone Group consists predominantly of red or red-brown mudstones which are assigned to the Manchester Marls. There is no indication of the existence of a basal sandstone unit that might represent the Collyhurst Sandstone of south Lancashire and north-east Cheshire (Smith et al., 1974), though localised occurrences cannot be ruled out. Minor developments of dolomite, gypsum and anhydrite are present, indicating an affinity with the St Bees Evaporites of north Lancashire and Cumbria. The Manchester Marls are assumed to be of Permian age, while the overlying Sherwood Sandstone and Mercia Mudstone groups are Triassic. The two last were formerly known as Bunter Sandstone and Keuper Marl respectively, names with misleading chronostratigraphical connotations that have for a decade or more been considered inappropriate (Warrington et al., 1980). The Sherwood Sandstone Group has not been subdivided, for although its presence has been proved in a large number of boreholes (Figure 14), these have not produced enough data to provide a basis for subdivision.

Knowledge of the stratigraphy of the Mercia Mudstone Group is largely dependent on recent work in adjacent districts, particularly around Blackpool (Wilson and Evans, 1990), where four formations have been recognised. The subdrift outcrops of these continue eastwards into the Garstang district. The formations comprise the Hambleton Mudstones at the base, overlain successively by the Singleton Mudstones, Kirkham Mudstones and Breckells Mudstones. These units were defined mainly on the basis of lithological distinctions recorded in boreholes drilled and logged in detail by the British Geological Survey. Details of the boreholes, together with data from the deep borehole

drilled by British Gas at Kirkham (SD 43 SW/6), are given by Wilson and Evans (1990). The Mercia Mudstones sequence shown in Figure 13 is based largely on that borehole.

The approximate chronostratigraphy of the Permo-Triassic succession is shown in Figure 13 (after Wilson and Evans, 1990, figs.3 and 14). The subdivision into stages was made by Dr G Warrington, based on his identification of palynomorphs in some of the boreholes in the Blackpool district and the Kirkham Borehole.

MANCHESTER MARLS

The basal Permo-Triassic formation, the Manchester Marls, is inferred to crop out beneath a thick drift cover in places between Garstang and Grimsargh (Figure 14). In the absence of evidence to the contrary, it is assumed that the outcrop here is not cut out by overlap of the overlying Sherwood Sandstone Group. The interpretation therefore differs from that adopted by Barker (1974, fig. 5), but is supported by borehole evidence showing incomplete sequences, 22 to 28 m thick, at outcrop (Figure 14). The westward thickening and downstepping by normal faults illustrated by Barker is supported by seismic evidence, however. The thickening is probably irregular, due partly to relief in the underlying surface of unconformity on Carboniferous strata, but mainly to movement on syndepositional faults during sedimentation. The latter effect is especially marked on the flanks of the Kirkham Basin, which is bounded on the west side by the Larbreck and Thistleton faults, and on the east side by the Woodsfold Fault (Figures 21 and 24).

The formation has been encountered in at least seventeen boreholes in the east, between Garstang and Fulwood (Figure 14). Of these, only two (SD 53 NW/4 and 53 SW/7) near Barton penetrated both top and bottom of the formation. The maximum thickness proved is 28 m.

Red or reddish brown mudstone is the dominant rock type proved in the boreholes, with subordinate beds of red sandstone in some sequences. Minor occurrences of gypsum are recorded from three of the boreholes. Borehole SD 54 SW/21 is exceptional for this area in containing 14.79 m of dolomite and dolomitic limestone, with some interbedded mudstone yielding ?Bakevellia. The presence of this bivalve supports the Permian age assigned to the sequence. Dolomite has been recorded in only one other borehole, SD 44 NE/13, just south of Forton (see below). These carbonate rocks probably equate with the Magnesian Limestone of south Cumbria, which Evans and Dunham (in Rose and Dunham, 1977, p.60) suggest formed around the margin of the Zechstein Sea (known as the Bakevellia Sea west of the Pennines). The concept was extended by Jackson et al. (1987), who envisaged a coastal sabkha as the environment of deposition.

The Manchester Marls are generally regarded as having been deposited in the Bakevellia Sea Basin (Smith et al., 1974) during late Permian times. There is no evidence of the

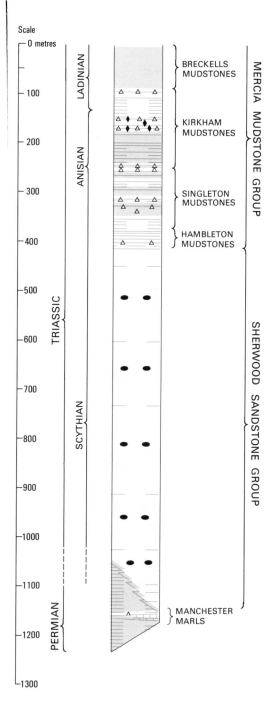

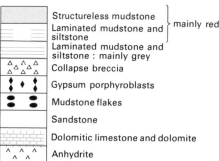

Figure 13 Generalised vertical section of the Permian and Triassic rocks.

eastward extent of this sea in the relatively low ground between Garstang and Longridge, but in the Ribble valley, to the south, an extensive elongate embayment is suggested by the presence of a small outlier of probable Manchester Marls near Great Mitton in the Clitheroe district (Earp et al., 1961, p.208). The main depocentre of this basin lay to the west under the Irish Sea where halite forms a major proportion of the sequence (Jackson et al., 1987, fig. 7(b)).

Details

The extent of borehole provings of the Manchester Marls is shown in Figure 14. Two boreholes near Barton, SD 53 NW/4 and SD 53 SW/7, show apparently full thicknesses of the formation of 4.88 m and 19.89 m respectively. The latter sequence is atypical in that the lowest 12.57 m consist of 'hard, red, marly sandstone'. Some details of these and other boreholes mentioned in the text are given in Appendix 1. The thickest sequence (28 m, top and base not proved) is in borehole SD 54 NW/3, where it is wholly argillaceous except for dispersed grains of coarse sand in the middle part and bands of gypsum up to 13 mm thick in the basal 5.8 m. There are two other records of gypsum in the sequence but the mineral is present only in the form of dispersed crystals in the mudstone in boreholes SD 43 SE/6 and 54 SW/2. In borehole SD 54 SW/6 sand grains dispersed in the red siltstone have a well-rounded ('millet seed') shape suggesting that they are aeolian grains blown into the sea from a nearby land area.

The dolomite (dolostone) and dolomitised limestone proved in borehole SD 54 SW/21 is varicoloured, ranging from red-brown to purplish yellow and pale grey. Colour mottling is common, as are pale vughs lined with dolomite crystals; one 0.91 m bed displays a honeycomb texture. Some beds show fine lamination with varying degrees of contortion, possibly representing algal mat lamination partially disrupted by solution brecciation. One thin section (E41084) shows simply a mosaic of finely crystalline dolomite, while another (E41085) from a more argillaceous dolosiltstone bed shows dolomite crystals in a muddy matrix, with scattered quartz grains and hematitised blebs. No recognisable bioclasts were observed in thin section, but one sample of core (NA 681) does contain a bivalve identified as *?Bakevellia* by J Pattison.

Dolomite was proved in only one other borehole (SD 44 NE/13), within a 3.6 m-thick Manchester Marls sequence. The dolomite bed is 0.30 m thick, pale yellow in colour and has a finely saccharoidal texture and anastomosing calcite veinlets; it lies 0.6 to 0.9 m below the junction with the Sherwood Sandstone Group.

SHERWOOD SANDSTONE GROUP

The Sherwood Sandstone Group in the district is undivided and ranges in thickness from about 450 to 1100 m. This is considerably less than the general thickness in the main part of the Irish Sea Basin to the west (Jackson et al., 1987), lending support to the view that the Fylde region lay on the margin of this basin. Within the present district, the thickness variations are mainly due to the effects of synsedimentary growth faults which gravity and seismic evidence (pp.86, 89) indicates are present at the margins of the Kirkham Basin (Figures 21 and 24). The bulk of the sand in this basin is thought to have been deposited by a great river system that flowed northwards from the northern France–English Channel region via the Worcester and Cheshire basins (Audley-Charles, 1970). In the Fylde region it is likely that there was also some sediment input from the

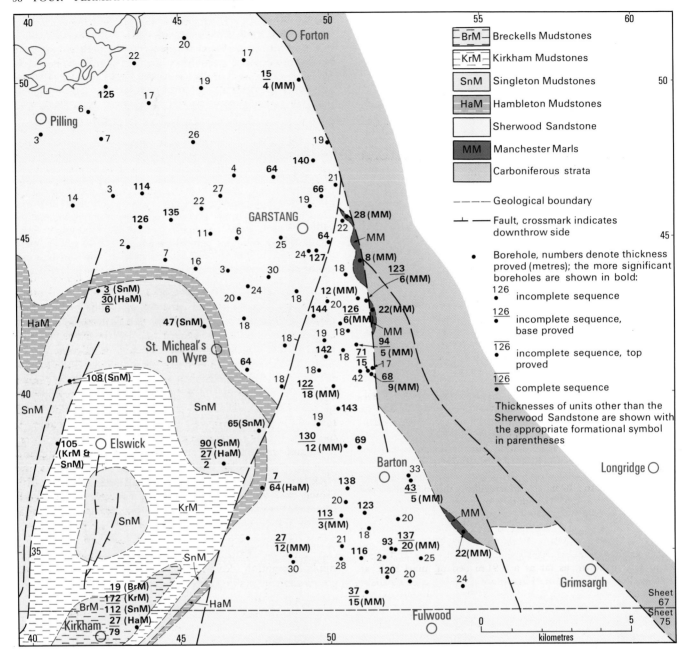

Figure 14 Thickness data for Permo-Triassic rock units proved in boreholes.

Carboniferous hinterland to the east, but evidence from provenance studies is lacking. Of the many boreholes drilled into the sub-drift outcrop (Figure 14), twenty prove over 100 m of Sherwood Sandstone, the maximum thickness proved being 144 m in SD 44 SE/29 near Catteral. Despite these numerous provings, lack of fossils and known marker bands precludes detailed correlation between sections.

The sandstone is generally red to red-brown and fine to medium grained, with a few coarser bands. Rounded quartzite pebbles, common in the English Midlands succession, are rare here. Partings or thin beds of red silty mudstone ('marl') are commonly present, and flakes or subangular clasts of similar mudstone are recorded at a few levels. Mudstone beds are rarely thicker than 0.6 m.

On a regional scale, the Sherwood Sandstone is generally regarded as being of mixed water-laid and aeolian origin (Jackson et al., 1987), with the former predominating in the lower part. In the district, the abundance of mudstone beds, at least in the basal 130 m of the sequence, supports this view. The lack of borehole to borehole correlation of these mudstone beds suggests that they were mostly deposited as overbank sediments.

The uppermost part of the Sherwood Sandstone is known in more detail from the Kirkham and Weeton Camp boreholes drilled in adjacent districts and described in detail by Wilson and Evans (1990). These authors distinguished three members: a lower one of mixed aeolian and fluviatile origin, a middle fluviatile one, and an upper one of dominantly aeolian origin. They note that these three facies are similar to those of the Helsby Sandstone in north Cheshire (Thompson, 1970; Earp and Taylor, 1986).

Details

The only good exposure of Sherwood Sandstone in the district is in a cutting [5046 4527] of the former Garstang to Knott End Railway, about 1.2 km east of Garstang, and was first described by De Rance (1877, fig.11). The exposure consists of 3.66 m of soft, red, medium-grained sandstone. Bedding is rather irregular, with some cross-bedding displayed in sets about 0.40 m thick. A water borehole (SD 54 NW/4), drilled in the floor of the cutting, proved red and brown sandstone to the bottom of the hole at 22.25 m, with a 0.92 m bed of reddish brown mudstone and siltstone at 13.11 m. The sequence probably lies in the basal part of the Sherwood Sandstone, for the underlying Manchester Marls are proved beneath drift in another borehole (SD 54 NW/3) sited only 200 m away to the north-east. The cutting was first described by De Rance (1877, fig.11), who referred to the sandstone there as 'Permian'. The precise age is not known, however, and the less precise term 'Permo-Triassic' would be more appropriate. The only other exposures are on the right bank of the River Wyre [4967 4856] near Scorton and in the River Brock [4929 4025] north of Myerscough Lodge. Both are small exposures of red sandstone and have little stratigraphical value. They demonstrate the thinness of the drift cover at the two localities, however, and may have important hydrogeological significance (p.93).

Numerous boreholes, drilled for water abstraction and monitoring, prove the presence of the Sherwood Sandstone (Figure 14). The logs have mostly been made by the drillers and provide little information of stratigraphical or sedimentological value. Many logs mention the presence of 'marl' bands, but the thickness and frequency of these vary considerably. For example, borehole SD 54 SW/5A, drilled at Stubbins, mentions eight such bands ranging in thickness from 0.2 to 1.82 m, spaced irregularly through mainly red sandstone in the lowest 55.17 m of the sequence. Another example, borehole SD 54 SW/9 drilled near Catteral, proved 126.34 m of sandstone from a depth of 13.87 m to the basal junction with the Manchester Marls at 420.21 m. Within the lower 97.54 m of this sequence, marl bands are recorded as being about 0.13 m thick every 1.22 to 1.52 m down, as far as a 0.91 m bed of 'marl rock' at 111.40 m. Below this bed, marl bands are recorded as being about 0.13 m thick every 0.91 to 1.22 m. A sequence of 'red marl' or reddish brown mudstone, with the exceptional thickness of 17.98 m, occurs 17.38 m above the assigned base of the Sherwood Sandstone in borehole SD 53 NW/5, north-west of Barton. The underlying sandstone in the lowest 17.38 m of the sequence is described as 'white', 'soft' and very fine grained, however, and the interpretation of the borehole record remains problematical.

Re-examination of some of the borehole cores by BGS geologists has revealed a few additional sedimentological details. These include the scattered presence of 'millet-seed' sand grains, indicating some aeolian deposition and, in borehole SD 44 SE/27, one red mudstone band containing desiccation cracks.

Three deep boreholes (SD 44 NW/2, 4 and 7) in an area about halfway between Pilling and St Michael's on Wyre are inferred from their geographical position to represent the middle part of the Sherwood Sandstone sequence. Only one of the logs (SD 44 NW/4) provides any detail, however; this record shows that the main difference between sequences at this level and those assumed to be lower in the succession is in the colour of the 'marl' beds, which tend to be grey or red and grey rather than predominantly red.

The borehole in the district that best illustrates the uppermost part of the Sherwood Sandstone sequence is SD 44 NE/2 near Woodsfold Bridge. The 64 m sequence, down from the junction with the Hambleton Mudstones at 42.67 m, is predominantly of fine- to coarse-grained red sandstone with a few grey beds and rare thin red 'marl' beds. A similar sequence, 79 m thick and logged in much more detail, is provided by the Kirkham Borehole (SD 43 SW/6), sited just south of the district boundary (Wilson and Evans, 1990). The reddish brown sandstone in the Woodsfold and Kirkham boreholes contrasts with the mainly grey colour of the uppermost Sherwood Sandstone sequence in the Weeton Camp Borehole, which was drilled just across the district boundary to the west [3888 3603]. The grey colouration here is attributed by Wilson and Evans (1990) to the effects of hydrocarbon leaching on the crestal part of the Weeton Anticline. The sequence is informally subdivided into three members on the presence of such sedimentary features as 'millet-seed' sand grains, adhesion ripples, cross-bedding, mudstone beds and desiccation cracks.

MERCIA MUDSTONE GROUP

Rocks of the Mercia Mudstone Group crop out beneath drift in the Kirkham Basin (Figure 21 and 24). The detailed stratigraphy relies mainly on the record of the deep borehole drilled for British Gas at Kirkham, just beyond the southern boundary of the district (Figure 14) and described in some detail in the Blackpool Memoir (Wilson and Evans, 1990). These authors have divided the sequence into four formations, primarily on the basis of their dominant colour (either grey, or reddish brown with bands of green in varying proportions) and on the presence or absence of well-developed lamination (Figure 13). Thus the Hambleton Mudstones at the base of the succession are dominantly grey interlaminated mudstones and siltstones; the overlying Singleton Mudstones are mostly reddish brown and structureless; the Kirkham Mudstones comprise mainly banded reddish brown and greenish grey mudstones interlaminated with siltstones, while the youngest formation, the Breckells Mudstones, is dominantly reddish brown and structureless. Other sedimentologically significant but stratigraphically less diagnostic characters occur, including ripple marks and ripple lamination, desiccation cracks, collapse breccias, salt pseudomorphs, halite veins and gypsum nodules, porphyroblasts and veins.

Salt beds form an important part of the Singleton and Kirkham mudstone formations in the Blackpool district to the west, notably in the Preesall saltfield. In the present district, however, salt beds as such are absent in the few boreholes which penetrate the Mercia Mudstone; however, some collapse breccias and pseudomorphs after halite are recorded, indicating that the mineral has been removed, presumably by groundwater solution.

The biostratigraphy of the Mercia Mudstone Group is described in detail by Warrington (in Wilson and Evans, 1990, pp.30–32), using fossil miospores (palynomorphs) extracted from the cores of boreholes drilled in the adjacent Blackpool district, and need not be repeated here. The conclusion, that the sediments were deposited mainly during the Anisian and Ladinian stages (Figure 13), almost certainly applies to the Garstang district also.

The fine-grained muddy sediments of the Mercia Mudstone mark a major environmental change from the coarser sandy deposits that accumulated during earlier Triassic times. Wilson and Evans (1990) envisage the grey

Hambleton Mudstones, with their alternation of subaqueous marine and evaporitic features, as representing repeated shallow marine incursions and retreats over an extensive flat coastal plain. The blocky structureless silty mudstones that occur commonly in the Singleton Formation and which form remarkably thick sequences in the Breckells Formation are thought to be of aeolian origin, presumably deposited directly from dust clouds. An additional or alternative explanation for the structureless character of the deposits is that they represent fossil soils, for recent research, summarised by Allen and Wright (1989), has shown that several processes may occur during pedogenesis, i.e. soil formation, that can destroy original sedimentary laminae. This may be indicated at some levels by laminae that show several stages of partial destruction (personal communication, Dr A A Wilson, 1989), the process having been arrested by renewed deposition.

Details

The full thickness of the Hambleton Mudstones has been proved in two boreholes in the district, SD 44 SW/2 at Crook Farm and SD 43 NE/1b at Inskip Creamery. At Crook Farm the formation comprises some 21.3 m of 'grey marl with gypsum bands', resting on 9.5 m of sandy 'red marl'; while at Inskip Creamery 27.1 m of grey 'marl' with sandy bands were recorded. In borehole SD 43 NE/2 at Woodsfold Bridge, 6.7 m of 'sandy marl' between the base of the drift and the top of the Sherwood Sandstone are tentatively assigned to the Hambleton Mudstones.

The Kirkham Borehole (SD 43 SW/6) proved 26.6 m of mainly grey and greenish grey, interlaminated mudstone and siltstone assigned to the formation, the siltstone being mostly in the basal 8 m, where it is mainly reddish brown. Collapse breccias and salt pseudomorphs occur in the lower part of the formation, whereas the upper part is characterised by cross-lamination, ripple marks and bands of mudflakes (Wilson, 1990).

The Singleton Mudstones in the Kirkham Borehole consist of alternating laminated and blocky facies. The base of the formation is taken at a depth of 339.60 m, above which the characteristic reddish brown colour becomes dominant. Interbedded collapse breccias are present in places; they may correlate with breccias and salt sequences proved in the Blackpool district to the west, such as the Rossall Salts and the Mythop Salts (Wilson and Evans, 1990, fig.6). Gypsum veins, generally 5 to 30 mm thick, are fairly common in the upper half of the formation, but are absent below 290.07 m. None of the available records of boreholes that were drilled within the district prove the full Singleton Mudstones sequence, the maximum thickness recorded being 108 m in the Wall Pool Bridge Borehole (SD 44 SW/1). The record of this borehole does not distinguish between the laminated and blocky facies, however, and the correlation made by Wilson and Evans (1990, p.13) is necessarily tentative.

The basal junction with the Hambleton Mudstones is recorded in two boreholes, SD 44 SW/2 at Crook Farm and SD 43 NE/1b at Inskip, at depths of 19.5 m and 112.8 m respectively, where the 'marl'

colour changes upwards from grey to predominantly red. The substantial thicknesses assigned to the formation at Inskip (89.9 m) and in two other boreholes in the district, SD 43 NE/4 (64.9 m) and SD 44 SE/11 (47.0 m), are logged mainly as red or brown 'marl' with gypsum, without additional significant detail. The only borehole recording the presence of salt and halite in the sequence is SD 43 NW/1, at Thistleton Bridge. An abridged log of this borehole is as follows:

	Thickness m
Drift	25.0
Marl, yellow-brown and grey	10.7
Marl, mainly red, with gypsum	57.7
Marl, green, gypsiferous, sandy	20.2
Marl, grey-green, with scattered halite crystals; 50 mm salt band near base	8.8
Marl, red and grey, with scattered halite crystals; 280 mm band of salt 178 mm above the base	6.53

The salt-bearing strata probably correlate with the Mythop Salts in the upper part of the Singleton Mudstones; higher beds probably belong to the Kirkham Mudstones (Thornton Member) but there is insufficient detail in the record to enable the boundary between the two formations to be drawn.

The Kirkham Mudstones crop out extensively beneath drift between Great Eccleston and Kirkham but, together with the Breckells Mudstones, are known in that area only from the Kirkham Borehole record (Wilson and Evans, 1990; Wilson, 1990). The Kirkham Mudstone Formation is readily divisible into three parts which have now been given member status; they are defined and described in the Blackpool Memoir and shown to be correlatable extensively in boreholes in that district.

In the Kirkham Borehole the base of the Kirkham Mudstones, and of the Thornton Member, is taken at a depth of 227.99 m, at the base of the lowest bed in the sequence, which shows strong reddish brown and greenish grey colour banding. Some brecciated and structureless beds also occur in the basal 8 m, above which laminated mudstones and siltstones predominate in alternating greenish grey and reddish brown units. There is some ripple lamination, and gypsum nodules and halite veins are fairly common. Blocky red mudstones with a few breccia beds again occur in the top 16 m. Structureless or faintly brecciated red mudstones with abundant veins and porphyroblasts of gypsum are present between depths of 111.18 m and 140.28 m; this sequence is taken to be the equivalent of the Preesall Salt Member. It is uncertain when the salt was removed by groundwater dissolution. The upper member of the formation, the Coats Walls Mudstones, proved from 68.38 m to 111.18 m, consists mainly of reddish brown, laminated mudstone with some structureless beds and breccias in the lower part.

The highest formation in the Mercia Mudstone of the Fylde region, the Breckells Mudstones, is inferred to crop out only in the central part of the Kirkham Syncline. It is present in the Kirkham Borehole from rockhead at 36.58 m down to 68.38 m and consists of red structureless mudstone with gypsum veins; brecciated beds predominate in the basal 13 m.   NA

FIVE

Quaternary deposits

Much of the district is covered by extensive spreads of drift. This can be divided into two major categories; glacial deposits, probably wholly of late-Devensian age, and postglacial deposits that were the result of diverse processes acting after the ice had melted some 13 000 to 14 500 BP (Gale, 1985). To date, no fossil or sedimentary evidence has been found for events predating the last glaciation; thus the early Quaternary history of the district remains uncertain.

ROCKHEAD

Though there are many boreholes into the Quaternary deposits, rockhead provings are scattered and many areas have no borehole coverage. Despite these limitations an attempt has been made to provide subdrift contours for the whole of the district. The resultant plot (Figure 15) purports to show only the regional pattern of rockhead gradients and should not be used for forecasting the depth to rockhead between data points. Greater detail is given in the area around and to the south of Garstang, where the depth to rockhead has been established by geophysical methods (Worthington, 1972).

If the interpretation of the point-data is broadly correct, then it is apparent that, were it not for the drift accumulations, the greater part of the Fylde and lower Ribble valley would lie below sea level. A notable feature of this area, where the Permo-Triassic rocks underlie the drift, is the network of deep, but narrow channels that radiate outwards from the edge of the Bowland Fells beneath the floodplains of the modern rivers. Two outlets to the west are tentatively identified, one beneath the River Wyre, the other following the line of Pilling Water. Many of the channels are overdeepened, with recorded rockhead reaching its lowest level of 36 m below OD at Fairfield Farm [4581 4220]. It seems likely that at least some of the channels were eroded subglacially, probably by meltwaters draining off the Bowland Fells. In other cases, meltstreams may have been diverted into existing pre-Devensian valleys.

The Permo-Triassic rockhead rises above sea level only locally, with rare surface outcrops; a series of low features can be traced in exposures from Little Crimbles, through Winmarleigh [468 469] to Manor House Farm, and a separate rise [415 349] underlies the M55 motorway. Whether such features have any geological significance or whether they are fortuitously preserved relics of a former topography is uncertain.

On the Carboniferous outcrop, postglacial erosion along the main river valleys has removed considerable thicknesses of drift, and the subdrift surface is unlikely to differ fundamentally from the present topography. The main uncertainty concerns the depth to rockhead through the Vale of Chipping, particularly at its western end. If, as proposed by Magraw (*in* Earp et al., 1961, p.240), the River Hodder once flowed to the Irish Sea by way of this valley, then a rockhead depression must exist beneath the watershed in the vicinity of Higher House Farm [593 383]. The presence of a subdrift channel in this position has yet to be substantiated.

GLACIAL DEPOSITS

During late Devensian times, ice lobes radiating from centres in the Lake District and adjoining Irish Sea Basin advanced across the lowlands in the west of the district. Subsidiary flows of Irish Sea ice, diverted around the Lancaster and Bowland Fells, may have merged with north Pennine ice in Ribblesdale before entering the district through the Ribble and Hodder valleys. The general pattern of movement, based on glacial striae, till fabrics and the orientation of drumlins and other ice-moulded landforms, is illustrated in Figure 16. At its maximum expansion (c.22 000 BP), the ice sheet probably covered the whole of the district, including the summits of the Bowland Fells, for Lake District erratics have been found at over 400 m on Parlick and glacial striae have been reported (Tiddeman, 1872) from similar heights on the Lancaster Fells.

The depositional products of the glaciation are dominated by till, which covers all but the most prominent bedrock features. The till is accompanied in the lowlands by sequences of outwash sediments forming multilayered complexes, in places up to 50 m thick. Downwasting of the ice sheet is believed to have occurred rapidly and was achieved largely by in-situ stagnation and decay of ice that had ceased to move. During this phase large volumes of meltwater were released subglacially, which eroded the Triassic rockhead surface to produce a network of channels. Later, as flow decreased, these became choked with gravelly outwash. Other meltwater deposits accumulated around the margins of the ice sheet and are now preserved as eskers and kames on the till surface. During the final stages of wasting, meltwaters were locally impounded to form transient glacial lakes.

The erosional effects of the ice are less apparent, but can be seen at the western end of the Bowland Fells, where the upland surface terminates abruptly in the truncated spurs of Nicky Nook and Harrisend Fell. Over-steepening of the north face of Beacon Fell, and possibly Longridge Fell, may also be attributed to erosive ice action.

Throughout the glacial episode, a periglacial climatic regime prevailed beyond the ice limits and solifluction was widespread. The resultant deposits, known as Head, form broad swathes on the higher slopes of the Bowland Fells above the mapped upper limit of the till sheet.

Till

The deposits of till comprise an ill-sorted mixture of rock fragments up to boulder size, set in a matrix of clayey sand or sandy clay. Their composition varies across the district,

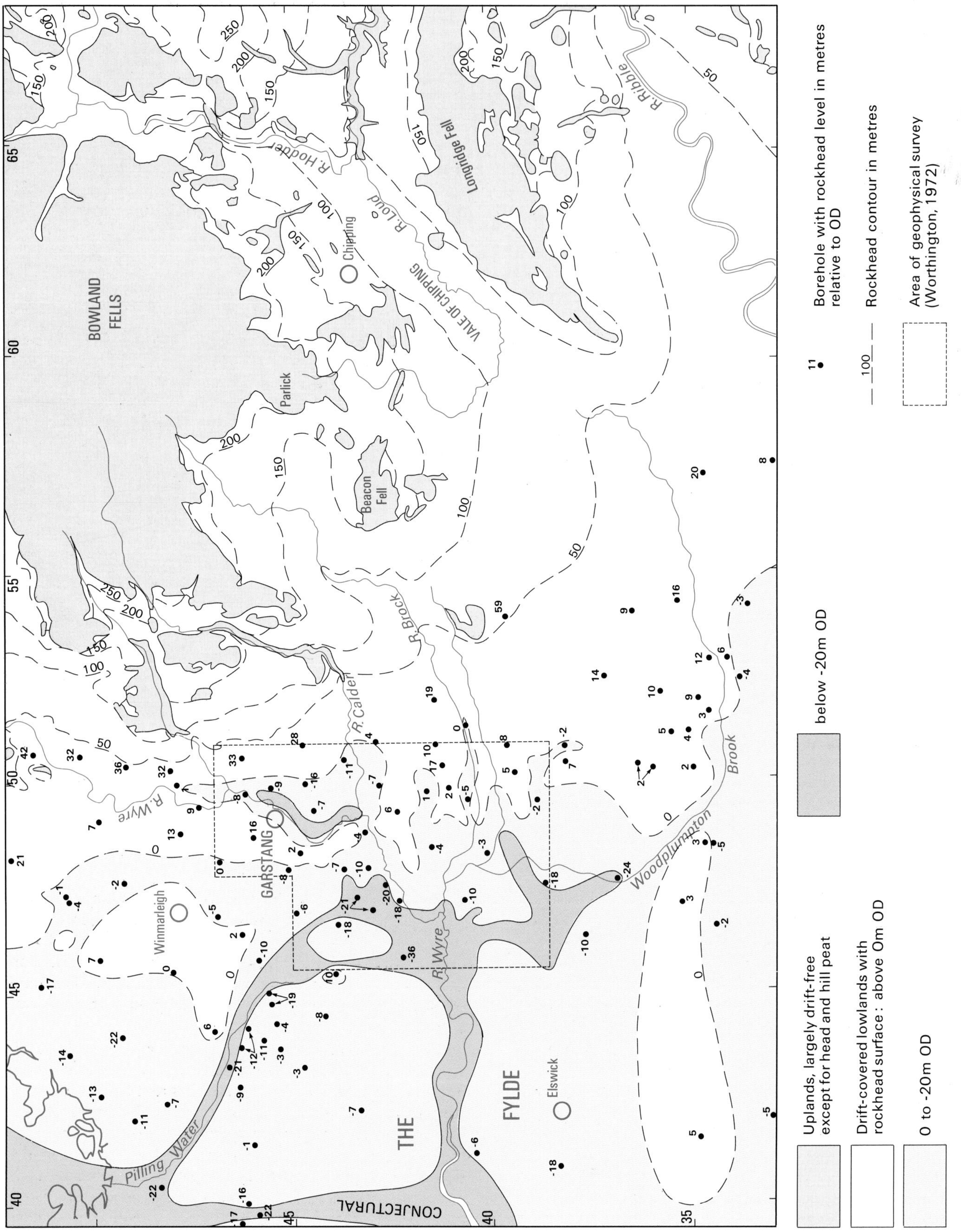

Figure 15 Subdrift bedrock topography of the district.

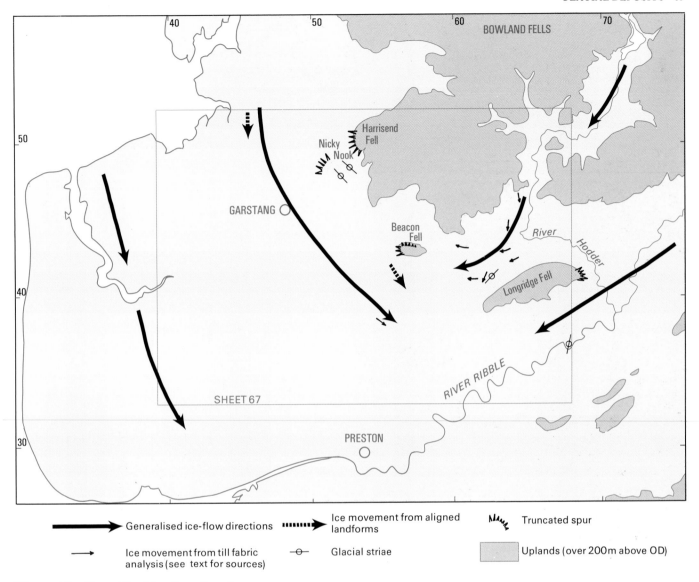

Figure 16 Generalised ice flow directions.

depending on the source of the ice, the nature of the local bedrock and the mode of deposition. In general, however, two broad till types are recognised: one is a red-brown till rich in debris of Triassic rocks, which was deposited by ice streams that crossed the Mercia Mudstone and Sherwood Sandstone outcrops of the Fylde and margins of the adjoining Irish Sea Basin; the other is a dark grey till rich in Carboniferous rock fragments and is essentially of northern or north-eastern provenance.

For description of the tills and their associated landforms, the district has been divided into three areas (Figure 17), each with distinct characteristics.

Area A (Eastern Fylde, Broughton, Goosnargh)

The broad tract of relatively low ground comprising Area A (Figure 17) extends across the eastern Fylde to the western foothills of the Bowland Fells. Farther south it is bounded by the watershed at the head of the Vale of Chipping and by the escarpment of Longridge Fell.

Throughout this area the drift cover is commonly between 20 and 25 m thick, but, exceptionally, thicknesses of over 50 m have been recorded, as shown by a borehole (SD 53 SE/12) south of Cow Hill. A marked thinning (to less than 10 m) occurs across high points on the Triassic rockhead and also against the higher ground formed by the Carboniferous outcrop.

The character of the glacial drift is best known in the south and east of the area, where many shallow boreholes were sunk to investigate the lines of the M55 and M6 motorways. The data are summarised in two generalised sections (Figure 18) . These show that for areas with a thin drift cover only a single till unit is recognised, but as the drift thickens so the sequence becomes increasingly complex and further subdivision of the drift is possible. Many of the deeper boreholes show, in upwards sequence: a basal till either in contact with, or close to rockhead; a number of impersistent buried tills associated with outwash sediments; and a widespread upper till sheet forming a surface capping. Although a broad

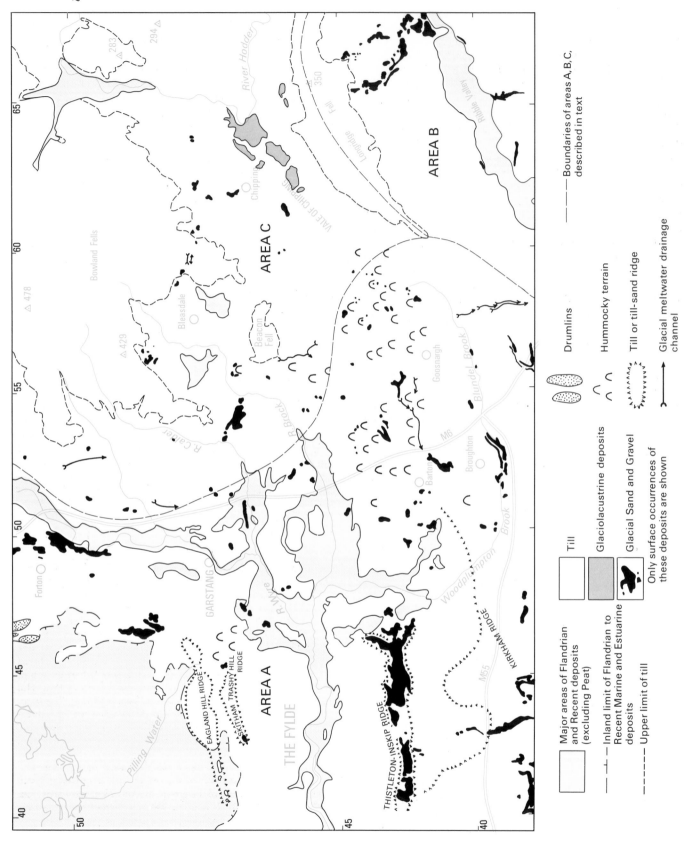

Figure 17 Distribution of glacial deposits and glacial landforms (postglacial deposits are shown

Legend:

Major areas of Flandrian and Recent deposits (excluding Peat)

Glaciolacustrine deposits

Glacial Sand and Gravel
Only surface occurrences of these deposits are shown

Till

— · — · — Inland limit of Flandrian to Recent Marine and Estuarine deposits

— — — — — Upper limit of till

— — — — — Boundaries of areas A, B, C, described in text

Drumlins

Hummocky terrain

Till or till-sand ridge

Glacial meltwater drainage channel

tripartite subdivision of the drift can be recognised locally, it is evident from a comparison of adjacent borehole logs that the sequence is more complex than the simple till–sand–till model propounded by De Rance (1877) and other Survey officers over a century ago.

Basal till

At the Corner Row intersection on the M55 (Figure 18, section A–B), boreholes for the bridgeworks encountered a basal till at 9 m above OD. The till is 2.7 m thick and described as a "hard, brown boulder clay". Drillers' logs refer to the difficulties of drilling this deposit by percussion methods and the geotechnical data confirm its high density and high cohesive strength. Few other boreholes on this motorway section reached bedrock, but several proved tills with similar mechanical properties lying clo-- e presumed base of the drift

Farth ...u, tne basal till crops out beneath glacial sands and gravels along the valley of Westfield Brook. Exposures in the valley floor [553 380] between Middleton House and Cross House reveal a dense, dark grey clay containing Carboniferous erratics of local derivation. Till fabric measurements (Longworth, *in* Johnson, 1985) indicate deposition by ice moving from the west-north-west.

Buried till

In the south of the area, lenses and sheets of till lie buried within outwash complexes. Some of the beds are up to 10 m thick and can be traced in boreholes for a kilometre or more, while others are lenticular bodies, probably only a few tens of metres across. In 1986, the working face in Bradley's Sand Pit [5111 3400] exposed two till layers within an otherwise uninterrupted sand sequence (Figure 19). Each comprised a flat-lying bed of stiff, reddish brown, pebbly clay streaked with sand laminae, and in basal contact with ripple-laminated, stone-free clay. The lithology of the till beds and their general disposition suggest that they may have been deposited subaqeously.

Upper till

An upper till sheet, 2 to 20 m thick, forms the main deposit mapped at surface across most of the area. The till was widely dug during the 18 and 19th centuries, when marling of the fields was practised as a means of improving land quality (Hall and Folland, 1970). The water-filled pits found in many fields are as a legacy of this era.

Minor and temporary exposures show that where the till is thickest it is a red-brown silty clay with abundant erratics, some of boulder size. The erratics, chiefly Lake District volcanics, greywackes and granites, form up to 70 per cent of the included clasts, but their numbers decrease eastwards (Longworth, *in* Johnson, 1985); other rock fragments are mostly Triassic and Carboniferous sandstone and siltstone. Thin beds of laminated clay, silt and gravel occur locally throughout the body of the till, and more substantial thicknesses of laminated clay are recorded in motorway boreholes to the west of Higher Bartle and in the vicinity of the M55/M6 interchange (Figure 18).

Around Forton, Carboniferous rocks lie close beneath the surface and, hereabouts, the Trias-rich till is replaced by a grey (weathering to yellow-brown) variant; the junction between the two till types is represented by a zone of interdigitation corresponding approximately to the mapped boundary between the Triassic and Carboniferous subcrops.

Data from the many motorway boreholes show that where the till is underlain by glacial sands it is draped across their surface as a conformable cover. Thus, hollows and ridges at the till/sand junction are broadly duplicated at ground surface (Figure 18). Longworth (*in* Johnson, 1985) has recorded postdepositional disturbances in the upper till sheet in motorway cuttings along the M55 and in trenches dug for the Catterall to Hoghton water pipeline. Slumps, folds and faults seen in these sections were ascribed to the in-situ melt-out of buried ice.

Details

Good sections in till are rare and tend to be confined to ditches and meander scars of the principal streams. Carboniferous-based grey till is seen, only in its weathered form, 30 m upstream of the M6/River Cocker crossing [4929 5375], just north of the boundary with the Lancaster district. Exposures of Trias-rich till are found in pits [4624 5146] dug in the drumlin to the east of Marsh Houses, and there are other occurrences around Elm Farm [4590 4620], at Kentucky Farm [4413 4618] and in the banks of the River Wyre, notably at Cross House [4851 4435] and downstream [4010 4128] towards Larbreck Hall. In all cases the deposit is the typical red-brown stony clay so common throughout the lowlands.

Between Elm Farm and Nateby Hall [4730 4625], cobble gravels, mostly consisting of sandstone, are concentrated into ridges on the till surface at about 9 m above OD. These may be lag deposits formed by tidal action at a time when the area was inundated by the sea.

In areas of slight topographic relief, discontinuous lenses of silt and sand are commonly encountered in the till. For example, in a stream section [433 438] through the till ridge at Ridgy Pool, silts, fine sands and pods of cobble gravel are closely associated with both stony and laminated stoneless clays. Other similar complexes are exposed to the east and west of Lousanna [4266 4497 and 4122 4464].

Landforms

The lowland tills give rise to three morphologically distinct landforms. These are drumlins, till ridges and tracts of hummocky terrain. Their distribution is illustrated schematically in Figure 17.

Drumlins, oriented north–south, form prominent features in a small tract of ground to the north of the River Cocker, part of a much more extensive drumlin field lying to the north in the Lancaster district. They are generally of large, being up to one kilometre in length, 400 m in breadth and up to 25 m high. Some have only a thin till cover and may be rock-cored. Farther south, there are isolated mounds of till of drumlinoid form at Wag Hill [4640 4325] and to the west of Black Hill Farm [439 457]. The latter are aligned east–west, normal to the direction of ice-flow as inferred from drumlins in the adjoining Preesall Field (Wilson and Evans, 1990), and their formation probably relates to ice stagnation rather than ice advance.

South of Pilling Water, the till plain is raised up in a series of parallel ridges, trending east or east-north-east. The Eagland Hill ridge, the most northerly of the group, extends 2 km eastwards from Birk's Farm [426 450]. It is up to 900 m wide and rises to almost 10 m above OD. The ridge is degraded and largely covered by peat at its western end, but small inliers of till at Lousanna and south of Hornby's Lane mark its position. A second ridge of till of similar proportions stretches from Skitham [4275 4362] to Trashy Hill [4485 4434], and a third, with an exposed sand core, extends from Thistleton [405 378] to Inskip [463 378]. The Kirkham ridge, the most southerly and highest of the group, lies to the south of

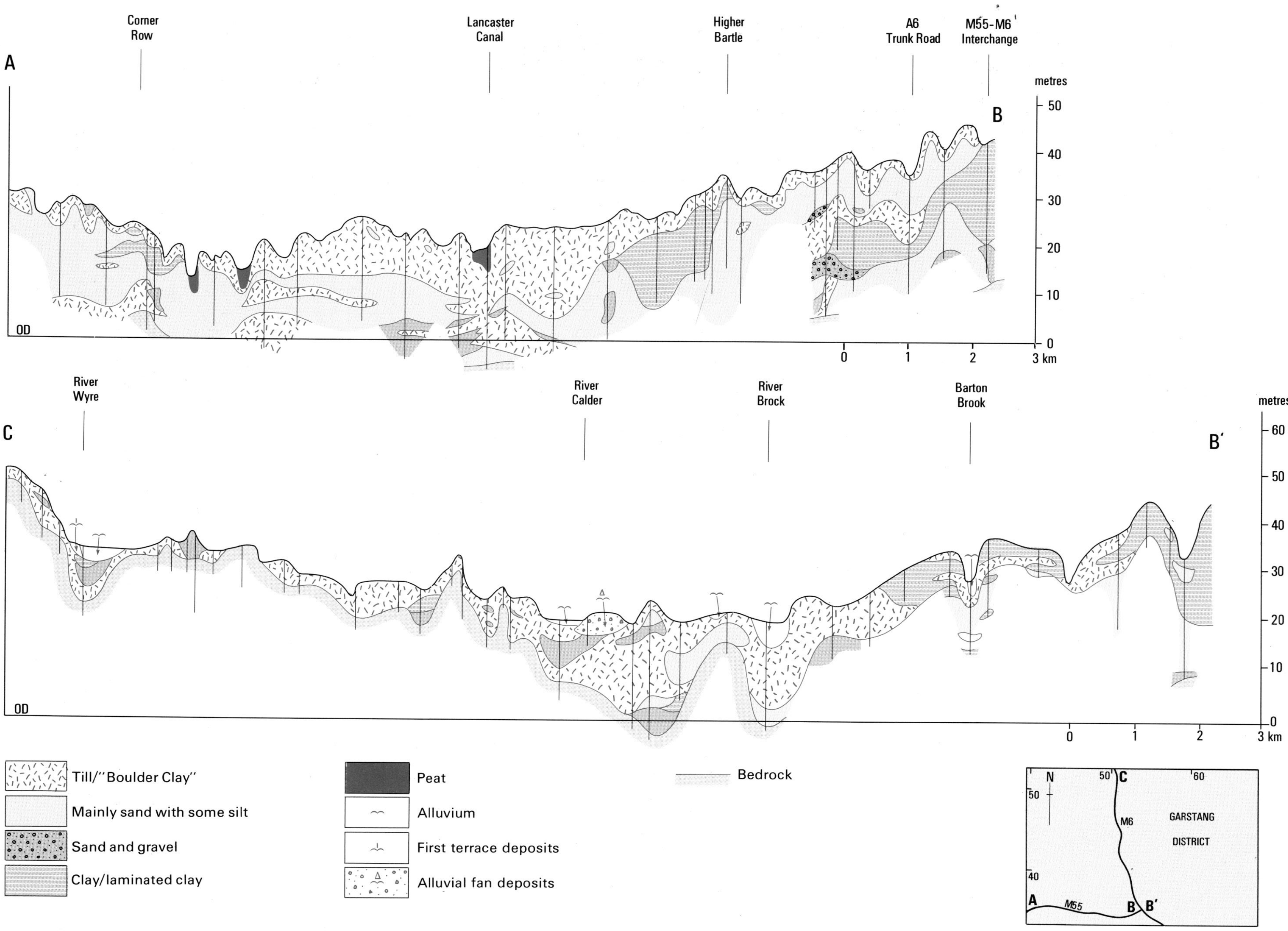

Figure 18 Generalised sections of superficial deposits along the M55 and M6 motorways.

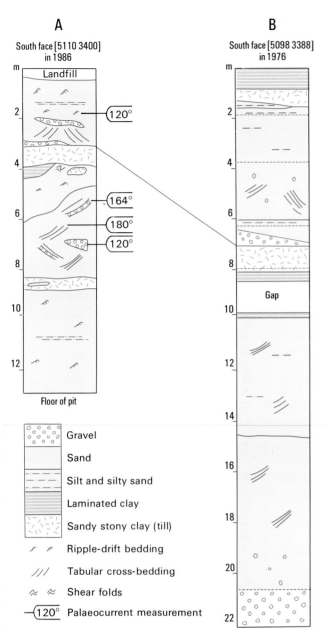

Figure 19 Generalised vertical sections in Bradley's Sand Pit.

and with diameters of between 50 and 200 m. Most show no strongly preferred orientation, but weak north-west and north-east trends are apparent. The mounds, which are composed of intercalations of till and sand, are well displayed on the ground [5285 3862] to the south and east of Green Lane Farm and [5320 3715] north-west of Barton House. Farther east, there are good examples between Cringlebrooks Farm [5790 3790] and Oak Tree [5860 3892]. A more isolated tract of hummocky, kettled terrain is found around Nateby Lodge [4616 4445].

AREA B (RIBBLE VALLEY)

From where it emerges from the rock gorge at Salesbury Hall [673 356], the River Ribble follows a deeply incised, meandering course through thick glacial drift. Till plateaux flank the valley to the north and south, but their smooth surfaces are in marked contrast to the more varied morphology of the lowland till plain.

The glacial drift is probably over 30 m thick south of the river [658 337] around Showley Hall, and nearer to 50 m thick on the north side of the valley around Grimsargh. The drift thins to the north and south against sandstone ridges in the Pendle and Warley Wise grits.

The till exposed at surface is a reddish brown sandy clay with grey mottling, similar in many respects to the upper till of the adjoining lowlands. It also, like its counterpart to the west, contains beds of laminated clay, sand and thin gravel. A darker coloured variant of the till can be distinguished close to rockhead, in which clasts of Carboniferous rocks predominate.

Price et al. (1963) accounted for the differing till types by suggesting that the Ribble valley was host initially to Ribblesdale ice, but was later invaded by Irish Sea ice. Longworth (*in* Johnson, 1985) expressed doubts about this reconstruction, and considered that the variation in the tills could be due to vertical differentiation within the ice rather than to two separate phases of ice movement. Quite how the various ice streams interacted in this part of the district is still unclear. However, there can be little doubt, given the northeasterly orientation of drumlins in the adjoining Clitheroe district (Earp et al., 1961), that the dominant ice flow was down the valley, towards the south-west.

Details

Sections in Park Brook [6755 3447] and in the deeply incised streams draining through Old Park Wood and Mercyfield Wood show a basal till comprising stiff, dark grey, stony clay with locally derived clasts of limestone, sandstone and mudstone. In the Park Brook section there is a progressive upward change from a basal till of Carboniferous provenance to a reddish brown till of more westerly Triassic derivation. Similar vertical differentiation is seen in landslip scars [6548 3322] west of Showley Hall.

The upper part of the till sheet was proved in a series of site investigation boreholes near Salisbury Farm on the outskirts of Grimsargh. A total of 61 boreholes, drilled over an area of about 0.5 km^2 and to depths ranging from 6 to 6.6 m, all recorded red-brown, sandy, silty, stony clay as the dominant lithology. In half of the boreholes, sand and gravel lenses varying between 0.3 and 3.7 m thick were encountered.

AREA C (BLEASDALE, VALE OF CHIPPING, HODDER VALLEY)

Thick spreads of till are present in the Vale of Chipping and the Hodder valley, and a generally more patchy cover exists

Woodplumpton Brook and forms part of a series of undulations that can be traced back through Kirkham [430 320] to Blackpool. First described by Gresswell (1967), this feature forms part of the so-called "Kirkham End Moraine". It has an irregular surface rising to about 35 m above OD and is dissected by closed, re-entrant valleys floored by alluvium or peat. The ridge is thought to have formed during a halt stage in the retreat of Irish Sea/Lake District ice. However, it is uncertain whether deposition occurred in a terrestrial, glaciolacustrine or glaciomarine setting. Several views are current (e.g. Johnson, *in* Johnson, 1985; Eyles and McCabe, 1989).

The irregular topography of the Kirkham ridge merges eastwards with a belt of more hummocky, kettled terrain which extends from Barton, through the headwaters of Westfield Brook, to the col at the head of the Vale of Chipping. Throughout this tract , the surface of the till sheet is moulded into low mounds, several metres in height around the foothills and lower summits of the Bowland Fells. The high moorlands are largely free of till, although small *remanié* deposits, less than a metre thick, may occur locally. On Parlick, till has been mapped up to a level of about 240 m above OD, but it reaches its maximum elevation of about 300 m on Marl Hill Moor.

The tills in the western sector of the Bowland Fells were probably deposited by ice streaming south-eastwards along the edge of the Bowland Fells; the glacial striae observed on the shores of Grizedale Lea [5320 4814] and Barnacre reservoirs [5237 4779] are consistent with this general pattern of movement. Farther east, measurements by Talbot (1974) of stone orientations from six sites in the Vale of Chipping indicate that ice from Ribblesdale was funnelled down the Hodder valley and was then diverted eastwards into the vale by the upstanding mass of Longridge Fell.

In the absence of boreholes, the thickness of the till can only be estimated where stream courses cut down through the till sheet to bedrock. In the Bleasdale area, thicknesses of between 4 and 14 m are common. Greater thicknesses (15 to 25 m), recorded in the incised gorges of the rivers Brock and Calder, probably infill pre-Devensian or subglacial valleys.

On the lower ground to the east of Beacon Fell and through the Vale of Chipping, the till cover gives rise to smooth fairly featureless slopes, the only exceptions being where it is moulded over dip and scarp features in the underlying bedrock as, for example, immediately north of Dairy Barn [633 436]. At least 18 m of till are present in the valley [619 437] above Chipping, and a similar thickness occurs in the Hodder valley [649 437], near Stakes.

The basal layers of the till sheet comprise a stiff, dark grey to grey-brown calcareous, silty clay or clayey silt with dispersed subangular to subrounded clasts, mostly pebble- or cobble-sized. The clasts consist mainly of Carboniferous rock types, which can be matched with local outcrops. For instance, the till around Chipping is particularly rich in Bowland Shale fragments, whereas, in the tills mantling the higher fellsides, blocks and boulders of locally derived Pendle Grit sandstone are the dominant rock fragments. Clasts of knoll-reef limestone are particularly distinctive in the tills around Cow Ark; they can be traced to outcrops of Clitheroe Limestone in the Hodder valley. Far-travelled erratics, usually present in small numbers, include Lake District andesites and hard, grey-green siltstones and greywackes that may have come from the Lower Palaeozoic outcrops in Upper Lonsdale. In weathered profiles, the till is less stiff in texture and has a paler, mottled grey-brown appearance due to decalcification. This process may also have removed many of the smaller limestone clasts from the top 1.0 to 1.5 m below the topsoil. In the west of the area, a distinctly redder facies of the till is found close to the Triassic subcrop, and similar reddening is observed in the tills that overlie the Pendle Grit outcrops north of Cow Ark. In the latter case, the reddening appears to be related to the incorporation in the till of large amounts of iron-stained sandstone.

Impersistent, but substantial bodies of sand and gravel may be present locally, concealed in the upper parts of the till sheet, but they are not generally distinguished on the maps. On the ground between the River Brock and Beacon Fell, they form a series of low mounds, some elongated and with a north-westerly trend, the highest rising 9 m above its im-mediate surroundings. Similar features can be observed roughly north-west of a line between Knot Hill [640 447] and Chipping.

DETAILS

Numerous exposures of till are to be found in the freshly eroded meander scars of rivers and streams in the area. Good sections are displayed along the course of the River Brock, especially upstream [5475 4373; 548 434] from Brockmill. At the latter locality there is an excellent section of an interbedded sand and gravel body, about 10 m thick, within till. Sharp contacts and upturned margins suggest an englacial channel fill. Englacial sands are also exposed higher upstream, particularly in the vicinity of Wood Top Farm [e.g. 5612 4426; 5604 4413], some 9 m being exposed at the first locality. Till forming the hummocky terrain to the west of Beacon Fell is unusually sandy or gravelly. A temporarary excavation into one of the larger mounds at Lower Barker [5534 4094] revealed the following section: topsoil, 0.33 m; on pale brown clayey sand, 0.55 m; on gravelly sandy clay with thin sand lenses, 2.25 m.

Farther east, the River Hodder and its tributaries provide many good sections. Unweathered slip faces can be examined along the valley of the River Dunsop and in gullies on the sides of the Hodder valley upstream from Dunsop Bridge [e.g. 6571 5105; 6638 5091]. Below the confluence of the rivers Hodder and Dunsop, small left-bank tributaries [6655 4878; 6620 4766] display up to 10 m of predominantly red stony clay. A sandy facies of the till is developed around Cow Ark [669 451 to 675 455].

Glacial Sand and Gravel

Glacial Sand and Gravel is more widespread than suggested by the mapping (Figure 17), because boreholes reveal that considerable thicknesses of sand, silt and gravel are present beneath a superficial covering of till (Figure 18). The principal deposits occur in three main settings: firstly, as channel-fillings overlying bedrock; secondly, as stratified spreads of generally cross-bedded sand within or beneath till; thirdly, as eskers and kames forming prominent constructional features on the till surface. In addition, there are more scattered outcrops of sand and gravel, which are difficult to classify; they are mostly of small size and are probably the result of sorting by supraglacial or englacial meltwaters. Such deposits are found commonly in areas of hummocky terrain. The main occurrences of each type of deposit are described below.

BURIED CHANNEL DEPOSITS

The buried channels beneath the eastern Fylde (Figure 15) contain variable thicknesses of sand and gravel, up to a maximum of 36 m proved in a borehole (SD 43 NE/2) near Woodsfold. In some channels, till forms part of the fill and may floor the channel, as shown by borehole SD 44 SE/24 adjacent to Pilling Water. More gravelly fill sequences are found in the upper reaches of the channels, close to the eastern margin of the lowland plain. This is well illustrated by the following log of a borehole (SD 44 NE/9) from the margin of the Wyre buried channel near Garstang:

	Depth
	m
Sandy gravel with cobbles	4.3
Coarse gravel with a few cobbles	6.1
Very coarse gravel	12.2
Fine gravel	19.8
Brown marl (?till)	21.3
Sherwood Sandstone	seen to 42.8

A similar sequence was proved beneath the River Brock, near Bilsborrow, where a deep borehole (SD 54 SW/13) encountered 1.83 m of silt and sandy clay, overlying 20.73 m of sandy gravel with minor clayey beds.

In the south-east of the district, sand and gravel covers a wide area between the River Ribble and Starling Brook. The deposits appear to infill a pre-glacial valley excavated by the River Ribble prior to its diversion through the gorge at Salesbury Hall. Landslip scars in the lower reaches of Starling Brook [6775 3679] and in the deeply incised valley [6809 3707] to the north show interbedded sands and gravels up to 20 m thick, overlying till. The deposits range from cobble-grade torrent gravels, with a high proportion of limestone and sandstone lithologies, to clean, free-running sands. Silty sands with associated laminated clays form a minor part of the sequence.

STRATIFIED SANDS

Stratified sands, formerly classified as Middle Sands (De Rance, 1877), lie beneath the upper till sheet in the south-west of the district. The sands attain their maximum thicknesses of over 20 m beneath the M55 motorway, where typically they are fine to medium grained and locally silty; in places, they are accompanied by gravel, with local lenses of silt, stoneless clay and "boulder clay". They thin rapidly to the north and east, and are not recognised to the north of the River Wyre, except as thin beds within till. The motorway boreholes testify to the discontinuous nature and variable thickness of these deposits (Figure 18), and show that it is difficult to correlate beds even between boreholes drilled for the same bridgeworks. Outcrops are restricted to the Thistleton–Inskip Ridge and to the valleys of the more deeply incised streams, notably Woodplumpton Brook, Barton Brook and Savick Brook. Better sections are displayed in Bradley's Sand Pit.

DETAILS

Thistleton, Inskip, Barton, Broughton

The largest spread of sand mapped in the district crops out on the Inskip–Thistleton ridge (Figure 17). Exposures are poor and are mainly confined to small diggings. At the western end of the ridge [4100 3759], 3 m of brown, fine- to medium-grained sand were seen in a disused pit. Farther to the east, near Inskip, excavations on the ridge top showed red-brown, mainly fine-grained sand, with beds of laminated sandy silt. The deposit was also formerly dug on a small-scale at Blue Moor. Boreholes sunk into the southern flank of the ridge at Inskip Creamery gave conflicting results, but in one well (SD 43 NE/1a) beds of gravel, sand, silt and clay, totalling 22.9 m, were reported. Although the mapping suggests that the sands generally pass beneath till, the relationships at the margin of the outcrop are complex, and in places a transition from sand, through silt, to till can be observed.

West of Beech Grove, the sands have been proved beneath till in three small pits [405 352], and there are more extensive outcrops in the deeper valleys to the south of the motorway.

At the time of survey, the sands were being actively worked at Bradley's Sand Pit [5110 3400], which exposed 13 m of beds beneath the till cover (Plate 14; Figure 19). Much of the working was in planar cross-bedded and ripple-drift bedded sand, but there were beds of till, laminated clay and channel gravels in parts of the section. The sand was generally fine to medium grained and clean, with many laminae in which finely comminuted shell debris and coal fragments were conspicuous, particularly in concentrations along foresets. Current directions, as indicated by the foresets, were mostly from the north and north-west. The sands crop out in the valley bottom immediately to the north of the pit, and for about 2.5 km upstream.

Other sections in sand are seen in the valley system of Barton Brook. For instance, a landslip scar [5235 3657] in the valley side north of Barton Hall showed 6.5 m of interbedded, cross-bedded sand and gravel, overlying a stiff brown stony clay, presumed to be a lodgement till. Springlines on the hillside south of the Hall mark the position of local clay or till beds within the main sand body. Higher upstream, discontinuous sands are exposed beneath till at Blake Hall [5377 3830], and also in the vicinity [5506 3784] of Middleton Hall. A section measured at the latter locality gave:

	Thickness
	m
Sandy gravel	0.8
Sand, red-brown, structureless	1.4
Sand, fining-upwards into clay, in sharp-based beds up to 0.03 m thick	0.2
Sand, red-brown, fine-grained, ripple-drift bedded, with some clay bands and carbonaceous concentrations	2.8

All the outcrops hereabouts show poor lateral continuity, and at no point do the sands appear to exceed 8 m in thickness.

More continuous bodies of sand, up to 7 m thick, crop out in the valley of Savick Brook, downstream [569 340] from near Clarkson's Fold. Exposures are restricted to bluffs and recent landslip scars where the brook has undercut its bank. The sand is orange, fine grained and ripple-drift bedded, with lenses of stony till and thin gravels near the base, and stoneless clay towards the top. In places, the sands can be traced laterally into stoneless clay or till. The sequence beneath the valley floor is proved by motorway bridgework boreholes at Hindley House (SD 53 SE/33 and 35). These encountered "boulder clay" in beds up to 7 m thick, with subordinate interbedded sands and gravels.

Ribble valley

Sands up to 5 m thick crop out in the steep-sided gullies that enter the Ribble valley from the south-east. The lateral continuity of these bodies is difficult to assess because of landslipping, but they appear to form flat-lying spreads at two levels, about 25 and 50 m above OD. A streamside cliff [6535 3396], 450 m north-west of Showley Hall, showed 4.7 m of orange sand sandwiched between tills.

ESKERS AND KAMES

Ridges and mounds of sand and gravel with the distinctive form of eskers and kames are a feature of the ground between White Carr and Cabus, and around Winmarleigh. Other sets of gravelly mounds can be traced down the dip slope of Longridge Fell to the River Ribble at Bailey Hall. Also placed in this category are low sand ridges that occupy a valley to the west of Grimsargh.

Plate 14 Till overlying sand at Bradley's Sand Pit [511 340] near Broughton. The trough cross-laminated, fine- to medium-grained sand is overlain by laminated clayey sand. Stony glacial till with a highly irregular base is exposed at the top of the face (A14790).

DETAILS

White Carr to the River Calder

Near White Carr [498 515; 495 515], steep-sided ridges with the classic form of eskers are constructed mainly of sand, with some gravel and cobbles. Another low, gravelly esker, just over a kilometre in length and up to 100 m wide, is present [507 436] north of Sturzaker House Farm. The linear, meandering form of this mound, lying subparallel to the River Calder, indicates that it was deposited by a glacial precursor of that river.

Another prominent ridge of sand and gravel extends along the west bank of the River Wyre, north of Cabus. Auger probings proved sandy clay and clayey sand, while an exposure [4941 4920] in the river bank south of Wyre Bridge showed 1.3 m of gravelly sand on 3 m of clay-free sand. Gresswell (1967) interpreted the ridge as an esker. However, as it is flanked on its eastern side by the River Wyre, it may be the remnant of a much more extensive kame deposit that has largely been removed by river erosion.

Sand and gravel, in part overlain by red-brown stony clay, forms a series of steep-sided mounds to the west of Winmarleigh. The deposits are lithologically very varied, with material from sand- to coarse gravel-grade. Several small pits, opened to exploit the deposit, are now degraded, but in one [4677 4733], poorly sorted gravel with thin beds of clayey sand was seen to a depth of 1.5 m. The pebbles comprised mainly Carboniferous sandstone with some Lake District erratics.

Ribble valley

Low, sinuous ridges of sand are preserved as constructional features in a valley [580 340] to the west of Grimsargh. The ridges, standing up to 2 m high, can be traced for about a kilometre before they disappear in the built-up ground adjacent to Longridge Road. They are interpreted as ice-contact deposits laid down by englacial or supraglacial streams draining towards the Ribble valley.

Ridges and mounds of sand and gravel rise above, and are presumed to overlie the till sheet on the northern flank of the Ribble valley. Two separate cross-valley systems are recognised. The more westerly one starts as a low symmetrical esker ridge on the outcrop of the Pendle Grit at Low Hill House [656 394], but degenerates south-eastwards into kame-like mounds, some standing between 10 and 30 m above the surrounding slopes. The other series of mounds crosses the escarpment to the east of Crowshaw House [666 394] and converges with the first in the vicinity of Bailey House [674 380]. A trial pit dug on the crest of one of the mounds [6761 3772] provided a bulk sample comprising: 27 per cent gravel, 25 per cent sand and 48 per cent clay. Almost the entire gravel fraction consisted of locally derived Pendle Grit sandstone.

Small deposits of glacial sand and gravel are found in areas of hummocky terrain. The main occurrences are roughly circular or ovoid mounds strung out along the eastern margin of the lowland plain between Blundel Brook and the River Calder. The largest spread [542 438], around Butt Hill and Foggs Farm, is mainly a pebbly clayey sand.

Glaciolacustrine deposits

In the Vale of Chipping, deposits of presumed glaciolacustrine origin, comprising soft sand and stiff, grey to red-brown, stone-free clay, underlie a tract of very low relief. This extends from the edge of the River Hodder gorge westwards as far as Hesketh Lane [620 412], interrupted only by the valleys of the Chipping [628 418] and Leagram [638 428] brooks. Though the sand and clay are separately delineated at surface, they are probably thinly interbedded at depth, as shown by a good exposure in the scar of an active landslip overlooking the River Hodder [6440 4345]. This provided the following section:

	Thickness
	m
Made ground	0.37
Sand, clayey, with scattered small pebbles	0.48
Gravel, sandy, consisting of small pebbles increasing in size downwards	0.54
Clay, reddish brown, stone-free, with indications of completely disrupted lamination	1.04
Sand, silt and clay, red-brown, interlaminated	0.34
Sand, pebbly	0.23

The lake, which was probably briefly established here after most of the Devensian ice had melted, would have covered an area of at least 4 km^2, with a surface level at about 103 m above OD. It would have ceased to exist once the River Hodder, below Stakes, and the River Loud had eroded their courses sufficiently to provide an outlet for the ponded water.

Contemporary glacial lakes may also have developed in the Hodder and Langden Brook valleys, above Whitewell, in a tract of ground now largely covered by river alluvium. Site investigation boreholes below landslips near Hodder Bank [6550 4838] have proved laminated clays over 9 m thick beneath the Hodder floodplain. There are also patches of intensely disrupted laminated clay exposed [6447 5051] beneath alluvium near Hareden.

Other deposits of purplish and grey stoneless clay are exposed [e.g. 6504 4182; 6575 4192] at the foot of Longridge Fell, notably between Bradley Hall and Head o' the Moor. These appear to be unrelated to the glaciolacustrine deposits noted in the Vale of Chipping, for they occur at heights of up to 140 m above OD.

Both Taylor (1961) and Gresswell (1967) have alluded to a major lake ("Lake Myerscough") which they believed once covered much of the lower Wyre catchment. According to

Gresswell, the lake may have been established in late-Devensian times; his reconstruction of the shoreline shows a body of water stretching from Eskham in the north to just south of Inskip. The field evidence for such a lake is inconclusive, because peat and modern alluvium now cover much of what would have been the lake bed. However, it seems unlikely, given the distribution of till, that a lake of the size envisaged by Gresswell ever existed as a single entity. Some support for local ponding comes from the reported occurrences of laminated clays in boreholes drilled near New Draught Bridge. Typical is the record of SD 44 SE/58, showing clay to 3.1 m, red-brown sand to 4.9 m, laminated silty clay to 6.7 m and sandy clay (unbottomed) to 8.0 m.

There are additional small outcrops of glaciolacustrine sediments [4435 4616] near Kentucky Farm, where Pilling Water cuts through a flat-bottomed depression and exposes crudely laminated, stoneless, red-brown clay and interbedded fine sand.

Glacial meltwater channels and late glacial drainage diversions

The distribution of meltwater channels recorded at surface during the present survey is shown in Figure 19. In the Bowland Fells two channels eroded in Pendle Grit cross the watershed between Parlick and Wolf Fell at a height of 400 m above OD. Farther west in the Bowland foothills, there are two channels, both incised in till and floored by thin patches of alluvium or head. One, to the north of New Hall Farm [509 459], can be traced northwards to Grizedale, from where there is a possible subdrift continuation through the valley between Nicky Nook and Harrisend Fell. The other, in the vicinity of Hill House [5723 4076], has a possible continuation along the valley from Ashes [5660 4093] east-south-east to near Loudscales, where it joins the River Loud.

Meltwater erosion may also have been an important factor in the formation of the lowland valleys, many of which are much wider and deeper than the present streams would have formed. This is particularly true of valleys such as Grizedale, which probably served as a meltwater outlet recovering drainage from the upper Wyre catchment. The misfit valleys of Barton Brook, Woodplumpton Brook and Savick Brook can also be cited in this context.

Late-glacial drainage diversions can be recognized in many of the main rivers. For instance, in the Ribble valley the gorge upstream of Salesbury Hall is a late-glacial or postglacial feature (Dean, 1950), for the old channel at this point lies to the north of the gorge beneath thick glacial drift. A windgap [681 369] high on the hillside north-east of the gorge marks another glacial derangement: Starling Brook originally flowed into Dean Brook through this windgap but was diverted southwards probably by late-glacial landslipping at the head of the dry valley (Dean, 1950). Since that time Starling Brook has deepened its valley by some 15 m.

The preglacial course of the River Hodder is conjectural. As noted earlier, the river may have flowed to the Irish Sea by way of the Vale of Chipping instead of following its present south-easterly course. If this were the case, the starting point must have been located in the present valley at Stakes, for immediately downstream, rockhead rises abruptly by at least 6 m. The narrow, drift-filled corridor, some 800 m wide, passing immediately south-east of Chipping and through Hesketh Lane marks the most likely course of such a pre-Devensian river.

Interpretation of the glacial drift

The origins of the stratified drift, of the type found in the lowland plain, have been the subject of vigorous debate for over a century. The assertion by Hull (1864) that the drift could be divided into a tripartite sequence of lower till, middle sands and upper till formed the basis for most early hypotheses. De Rance (1877) believed the lower and upper tills to be the product of glaciomarine sedimentation involving sea ice. He interpreted the intervening sands as tidal accumulations deposited around the coast of a gradually deepening sea during a period of climatic amelioration. Tiddeman (1872) rejected this hypothesis, favouring instead a model involving only land-based ice. In this and many later reconstructions successive tills were assumed to have been deposited during separate phases of glacier advance and retreat. In recent years, as the role of supraglacial processes has become better understood, multi-till sequences have been interpreted increasingly in terms of a single ice advance involving lodgement, melt-out and flow till elements (e.g. Boulton, 1972). Longworth (in Johnson, 1985) has described how such a model might be applied to explain the drift sequences of the Fylde.

Although none of the ideas that has emerged over the years fully explains the problems of the lowland drift, some general conclusions can be drawn. First, there can be little doubt that a basal, overconsolidated till of lodgement type underlies much of the district. In areas where the till cover is thin, this till probably forms the main glacial deposit. The geotechnical data from the motorway boreholes show that where succeeding tills are present they are weaker and less compact, and are, therefore, unlikely to represent readvance tills.

The origin of the succeeding stratified sands is more speculative. Their distribution in morainic ridges, such as the " Kirkham End Moraine", may be explained using the model suggested by Boulton and Paul (1976, p.165). In their view, sands are laid down by outwash streams flowing between ice-cored ridges. As the dead ice melts the topography reverses, leaving the sands standing as kames. There are at least two objections to this model. First, wherever palaeocurrent directions in the sands have been measured, they show flow from the north-west, not along the axes of the ridges as might be expected. Second, the model, as conceived, would require the upper till to be essentially a flow till. This seems unlikely, for although there is some evidence for mobilisation and flowage, the upper till mostly forms a consistent mantle that shows no thinning over underlying sand ridges.

In an attempt to resolve this problem in the adjoining Blackpool district, Wilson and Evans (1990) have postulated that the "Middle Sands" were deposited beneath floating ice; the upper till would then represent an ablation moraine deposited as the ice gradually melted and the water drained away. This view has particular appeal because it solves the problem of maintaining an ice cover whilst, at the same time, providing space to introduce substantial thicknesses of waterlain sediment subglacially. Implicit in this model is the assumption that decoupling of the ice sheet from its bed was caused by ponding of glacial meltwater. However, as discussed in recent work by Eyles and McCabe (1989), deglaciation may have occurred under conditions of high relative sea level. Under these circumstances, at least part of the lowland stratified drift may be of glaciomarine origin. The authors cite the Kirkham ridge as an example of a glaciomarine morainal bank formed where the ice margin stabilised in shallow water. Until further research is carried out, the distinction between possible glaciomarine sediments and other water-laid glacial deposits remains open to speculation.

It seems likely that in the final stages of collapse of the ice sheet large areas of stranded, dead ice wasted down in situ. This is confirmed by the widespread development of hummocky and kettled terrain and the prevalence of waterlaid outwash in the upper levels of the drift sequence. It is notable that the distribution of the hummocky terrain is mainly located in zones where opposing ice streams may have been in contact. Whether movement of the ice was checked in these areas and this contributed to ice stagnation remains uncertain.

POSTGLACIAL DEPOSITS

The early postglacial history of the district involved terrace incision, and accumulation of peat in kettleholes. Most of the head deposits and some of the landslips may also date from this period. Later, following a relative rise in sea level in mid-Flandrian times, marine and estuarine silts and clays were deposited on low-lying ground in the north-west of the district. The formation of peat "mosses" and alluvium on the lowlands, and hill peat on the moors, complete the Quaternary history of the district.

Head

Head consists of weathered near-surface bedrock or drift material that has become mobilised during repeated episodes of freezing and thawing, and caused to move downslope by solifluction processes. Most of the head probably dates from the period immediately following retreat of the ice, when cold periglacial conditions would still have obtained; however, some of the deposits may relate to earlier cold stages. In more recent times, further downslope surface movement and concentration of such debris has taken place, mainly by gravitational creep and downwashing.

Head is present on all sloping ground, but has been mapped only where its thickness is considered to be consistently greater than one metre. Slope deposits derived from bedrock form extensive sheets, more especially in the Bowland Fells, where they overlap the upper margins of the till sheet. The accumulations typically show surface lobes indicative of gravity flow. Head also fills deep stream incisions and occurs at the foot of over-steepened rock scars. In places the deposits have been reworked and incorporated in fluvial deposits and landslips. Soliflucted drift deposits are probably widespread on the lower slopes, but are not readily distinguished from their parent material and are normally included with it for mapping purposes. Head varies widely in composition, but on the Pendle Grit outcrops it typically consists of a poorly consolidated sandy clay with ill-sorted, commonly angular sandstone fragments. The thickest head occurs in the north-east of the district where in some exposures it exceeds 10 m; elsewhere, it is generally less than 3 m thick.

Details

Eastern Bowland Fells and Longridge Fell

Head deposits are widespread on the steep valley sides along the River Hodder and its tributaries. In the Langden valley, head consisting almost entirely of unsorted sandstone fragments attains a thickness of 4 m. At one time the deposits probably covered the entire valley floor, but today they are being actively eroded by Langden Brook.

Aprons of head form lobe-like spreads up to 12 m thick on the western flanks of Hodder Bank Fell and on many of the steeper slopes along the Hodder valley. In places, below rock scars, scree has been incorporated into the head deposits, as for example above the Hodder gorge [6540 4655]. Many of the outcrops hereabouts are inherently unstable and are intimately linked with small landslips.

On the north-facing escarpment of Longridge Fell, outcrops of head consist of up to about 2.5 m of sandy clay or clayey sand with unsorted or crudely stratified angular rock fragments, mostly of Pendle Grit or Pendleside Sandstone. Relatively undisturbed head deposits are limited to a small area [652 403] on the fell top, where they comprise some 3 m of sandstone-rich clay, in places leached to a whitish colour; the southern margin of the deposit is partly soliflucted over chert-bearing cobbly till.

Western Bowland Fells and Beacon Fell

Head deposits, up to 3 m thick, mantle the north-facing slopes of Catshaw Fell, and are exposed in Catshaw Greave; they mainly comprise soliflucted sandstone debris. Farther south, on Nicky Nook, patches of head, mapped in rushy hollows, are probably soliflucted till. Much of the drift hereabouts has probably moved to some extent but it is not sufficiently distinguishable from the parent material to be separated as head. However, an area on the steep slopes of the valley of the River Calder [529 443], north-east of Sullom End, has been delineated; there are no exposures.

Head consisting of clay with sparsely dispersed small stones infills the glacial meltwater channel [561 417] between Oaken Head and Crombleholme Fold. Nearby, head has accumulated on the lower part of the steep north slope of Beacon Fell; a gully exposure [5677 4316] revealed some 3.5 m of poorly sorted mudstone, siltstone and sandstone debris in a brown clayey matrix.

Flandrian Marine and Estuarine Deposits

An almost flat spread of estuarine and shoreline deposits, classified as Older Marine and Estuarine Alluvium, covers much of the north-west of the district, where the postglacial sea has invaded low-lying parts of the glacial drift surface. Deposition probably commenced some 7000 to 8000 years BP (Tooley, 1980) following a rapid rise in sea level and a landward expansion of a 'proto'-Morecambe Bay. Gresswell (1967) has argued that by 6000 years BP sea level was some 5.7 m higher than at present, and the coastline ("Hillhouse Coastline") lay several kilometres inland. Near Winmarleigh its position was identified from a line of low cliffs cut in till. Tooley (1980), working in the south of the Fylde,

refuted the concept of a Hillhouse Coastline and has argued convincingly that restoration of sea level during Flandrian times was achieved by a succession of marine transgressions separated by regressive phases. To what extent this theory can be applied in the present district is uncertain, as the sedimentary record is only sketchily known from drillers' logs, and peats signifying regressive stages have not been widely recorded.

The main body of estuarine deposits lies more than 6 m below OD and covers an area of some 30 km^2, stretching inland for up to 5 km. Crossing its surface are meandering depressions marking the position of former tidal creeks. On Cockerham Marsh, many of these creeks have now been filled with sandy material won from the foreshore. Near the coast, the outcrop is covered by patchy deposits of fine sand, up to a metre thick, that may be, in part, wind-blown. Farther inland, and covering much of the inner margin of the deposits, there are extensive areas of peat. Much of the area lies within the range of the highest tides and would be subject to flooding were it not for the protection afforded by the sea wall constructed in the 17th century. This barrier is taken as the arbitrary boundary between the Marine and Estuarine Alluvium and the Older Marine and Estuarine alluvium. Currently, sediments are accreting on the seaward side of a new barrier constructed in the last decade and shown on the 1:50 000 map. The sequence is up to 22 m thick beneath the present coastline but thins progressively inland to a feather edge against a rising till surface. The deposits proved in boreholes comprise mainly grey and grey-brown silts and clays, with sands, gravels and shell-beds. A thin peat has been drilled at the base of the sequence at Pilling. Low gravel ridges rise above the general level of the estuarine deposits close to their inner edge; they are interpreted as storm beaches, marking the approximate landward limit of marine transgression. They are classified on the map as Older Storm Beach Deposits.

Contemporary intertidal deposits, consisting mainly of silt and mud with a little sand, are accreting in the tidal flat area beyond the present sea-wall. The sediments are gradually being colonised and stabilised by salt-tolerant plants.

Details

There are many exposures of the older estuarine deposits in the shallow drainage ditches that cross the area. The greatest thicknesses, however, are seen in the water courses of Pilling Water and Ridgy Pool, and the main left-bank tributary of the latter. Up to 3 m of structureless grey-brown silt was seen in Pilling Water [4155 4757] near Stake Pool. Similar thicknesses could be seen farther upstream [4324 4675] near Bone Hill Farm, and in Ridgy Pool [4127 4615] near Bradshaw Lane Head. Although many of the deposits appear structureless, a silty lamination is commonly seen in the sediments exposed in the tidal creeks close to the present-day shoreline. Locally, sediment of fine-sand grade is present at surface, and these somewhat coarser deposits are particularly notable around Little Crimbles and Shepherd's Farm. The thickest records of this deposit are in boreholes SD 44 NW/9 and 12 near Moss Edge and at Pilling, where the base of the deposit lies at about 16 m below OD. The junction with the underlying till is generally one of gentle overlap, as can be seen in Pilling Water [4525 4555] and in a ditch section [4534 4629] west of Elm Farm. However, from Great Crimbles to some 400 m south of Hardhead, the outcrop of estuarine deposits terminates abruptly against a sharply rising till surface. Over the northern part of this stretch, a small cliff has been eroded into the till, perhaps by wave action.

Low ridges of gravel, thought to represent remnants of marine beaches, rise above the general surface of the deposit at several localities. The best preserved is centred on Cogie Hill [4422 4686] and consists of a sinuous ridge of gravel up to a metre high and 30 m wide. It trends south-west and can be traced for almost 2 km. The ridge bifurcates at its eastern end; the northern branch dies out and the trend is taken up by a new ridge 100 m to the south of the southern branch. West of Cogie Hill the ridge is far less conspicuous, but is well exposed [4343 4666] where it crosses Pilling Water. Here it is seen to comprise rounded to well-rounded gravel with medium to coarse sand. The gravel includes a large proportion of Lake District erratics, the remainder being made up of grey Carboniferous sandstone.

Other, shorter gravel ridges are found near the eastern margin of the deposit, north of Great Crimbles. At two localities [4631 5112; 4550 5070] the ridges are quite low and the gravel core was only proved by augering.

River Terrace Deposits

Terrace deposits are associated with all the major rivers and streams but are best developed in the Hodder valley upstream of Whitewell and along the winding valley of the River Ribble. Some of the higher terraces may represent fluvioglacial aggradations, dating from late-Devensian times, but most are Flandrian in age. The deposits typically consist of upward-fining and laterally graded spreads of gravel, sand, silt and clay.

Details

River Hodder and tributaries

Terrace deposits form extensive spreads in the lower Dunsop and Langden valleys, and along the main Hodder valley between Dunsop Bridge and Burholme Bridge. The terraces,which are mainly founded on till, grade to a nick point located at the head of the Hodder gorge south of Whitewell. The precise correlation of the terrace fragments, particularly in the vicinity of Burholme, is unresolved. Harvey and Renwick (1987) recognised three terrace levels hereabouts, with surfaces at 8 m, 4 to 5 m, and 2 to 3 m above the floodplain. The highest terrace, they suggested, was a late-glacial aggradation, whereas the youngest gave a radiocarbon date of c.5000 years BP or earlier. In the present survey, only First and Second Terrace deposits have been widely mapped, though a lower set of younger terrace fragments can be recognised locally.

Flats classified as Second Terrace are best developed on the right bank of the River Hodder, west of Langden Bridge. They consist of a metre or so of gravel, overlain by 3 m of silty clay containing mudstone fragments. The latter are probably derived from scree material reworked from the adjacent Bowland Shales outcrop. The front scarp hereabouts is very prominent, and alluvial fans, landslips and downwashed head characterise its length. Other Second Terrace occurrences are generally thin and patchy, and only partially cover erosional benches cut in till.

Spreads of First Terrace deposits are well represented on either side of Langden Brook near its confluence with the River Hodder, and in the main valley opposite Burholme. In the terrace-flat on the north side of Langden Brook, up to 2 m of cobbly gravel and gravelly sand are exposed in landslips [e.g. 6595 4997]. On the opposite bank, a section [6556 4976] through the front scarp showed in upwards succession; 3 m of till, 1.5 m of coarse gravel and 0.4 m of silt. The deposits have been dug at several localities nearby.

Lying between the First Terrace and the floodplain are small terrace renmnants, collectively grouped as Undifferentiated Terrace Deposits; a stack of a least four [664 499] just south of Thorneyholme consists mainly of coarse gravel and surface silt.

In the River Loud, isolated terrace fragments rising about 2 m above the alluvial plain are present between Holwood [594 398] and near [634 421] Gibbon Bridge.

River Ribble

The distribution of terraces in the Ribble valley relates to the blockage of the river by thick glacial drift in the adjacent Clitheroe district and its subsequent diversion upstream of Salesbury Hall through a gorge cut in Warley Wise Grit. Thin veneers of brown silty clay, designated as Third Terrace, cap gentle slopes of till on the northern outskirts of Ribchester. The village itself stands on deposits of the Second Terrace, which rise to between 3 and 4 m above the floodplain. Over the years, lateral erosion by the river has destroyed part of this terrace level and with it a corner of the Roman fort that once stood on the river bank close to the village. On the left bank of the river, smaller areas of Second Terrace deposits have been mapped adjacent to Sunderland Hall [6268 3367] and also by Sheep Fold [6132 3290]. Extensive spreads of First Terrace lie within the meander loops and rise to 2 m above the Ribble alluvium. The terrace surface is undulating and gullied, and, in places, is difficult to distinguish from the modern floodplain.

River Wyre and tributaries

Two terrace-flats, lying at 1.0 and 2.5 m above the modern alluvium, are recognised in the Wyre valley, north of Shireshead. A borehole (SD 55 SW/9) sunk through the upper terrace proved 1.8 m of gravel, on 1.1 m of laminated clay resting on bedrock. Gravel workings have greatly reduced the size of the terraces and only fragmentary deposits remain.

Gravel and silt deposits form terraces up to 5.5 m above the present beds of the rivers Calder and Brock, especially the latter, as below Walmsley Bridge [5334 4141]. This particular terrace broadens greatly below the nearby gorge where there are extensive gravel workings. Its surface grades westwards at a slightly greater inclination than the present river course so that it eventually passes below the level of the alluvial plain. This relationship suggests that sea level may have been lower when the terrace was formed.

In Westfield Brook, terrace remnants rise to about 5 m above river level. They are composed of coarse, matrix-supported gravel up to 1m thick, overlain by about 1.5 m of grey-brown sand. The angularity of the clasts in the gravel and the preservation of glacial striations on the coarser boulders, have been interpreted by Longworth (in Johnson, 1985) as evidence that the gravel is of fluvioglacial origin. In support of this argument, he has shown that the base of the gravel in the north bank of the stream [550 378] is let down locally into subvertical-sided graben structures developed in the underlying glacial sands. For the graben to have survived, he argues, the sand must have been frozen when settlement occurred.

Alluvial Fan Deposits

Many of the streams flowing off the Bowland Fells have deposited alluvial fans at the break in slope with the lowland plain. Other smaller fans and cones are present where small, steeply graded tributaries join main rivers. The fans typically consist of poorly sorted gravel and cobble beds, with associated channel-fill sands and interbedded organic deposits.

A major alluvial fan has been deposited by the River Calder where it emerges from a gorge at Sandholme Mill, and another is centred to the south-east of Chipping. Similar, though less extensive fans flank the River Loud to the east of Blacksticks [599 435; 595 430; 599 420] and occur below ravines [583 451; 616 464] to the west of Parlick and north-west of Burnslack respectively.

There are many examples of small fans and cones at points where streams enter main valleys. The best are along Langden Brook and in the valleys of the Hodder and Ribble. Harvey and Renwick (1987) have demonstrated two major periods of fan aggradation affecting sites in the Langden valley. The first, after 5400 years BP but before 1900 years BP, overlaps with a period of marked climatic deterioration at about 2500 years BP (Lamb, 1982); the second, at around 900 years BP, occurred at a time when vegetational alteration could have been influenced by human agency. The larger fans probably have a long but intermittent history of sedimentation, with some commencing at the final stages of ice melt.

Details

The alluvial fan deposited by the River Calder was worked during the construction of the adjacent M6 motorway, and a lake now occupies the former borrow pit. Small exposures in the banks of the pit indicate a deposit rich in gravel but containing a high content of red-brown clayey silt. Site investigation boreholes associated with motorway construction show interbedded sand, gravel and silty clay to depths of 9.75 m; one (SD 54 SW/29) proved a 0.61 m interbedded peat layer at a depth of 4.27 m.

The fan centred to the south-east of Chipping [629 431] is clearly formed by sediment that has fanned out from the deeply incised valley above the village. It is possible that a flash flood originating from this valley occurred in 1851, when below Chipping were "many acres of land covered with mud and gravel, and strewn with wreck of every description" (Weld, 1851). Exposures in ditches [e.g. 6283 4268; 6331 4332] south of Dairy Barn indicate that the fan consists of gravelly sandy clay or clayey sand and gravel. The maximum thickness is unlikely to be much in excess of 3.0 m.

Alluvium

Alluvial deposits occur in all but the smallest valleys, and areas of lacustrine alluvium occupy local hollows in the till surface. The most extensive alluvial flats are those in the lower Wyre catchment between St Michael's on Wyre, Churchtown and Bilsborrow. Other major spreads occur in the Wyre valley north of Garstang and in the valleys of the Hodder and Ribble. The alluvium typically consists of an upper layer of sand or silt overlying a basal gravel. Thicknesses are normally in the range of 2 to 5 m, though 15 m of alluvium has been reported from one borehole (SD 44 SE/5) in the lower Wyre.

Details

River Wyre

A tract of alluvium up to 600 m wide has been deposited by the River Wyre where it flows southwards along the margins of the Bowland Fells. Exposures are poor, but boreholes (e.g. SD 44 NE/7) reveal a thin veneer of silt or silty clay over sands and gravels. The base of the alluvium is not easily determined because substantial thicknesses of sand and gravel, probably of fluvioglacial origin, underlie the valley. Gravel has been worked from beneath the alluvium [499 498] to the north-east of Wyre Bridge.

Below Churchtown, in an area of the Wyre catchment that is prone to flooding, the alluvial tract is up to 3 km wide. The deposits consist mainly of sandy silt between 2 and 3 m in thickness. Unusually, borehole SD 44 SE/5 recorded silts and silty clays, regarded as alluvium, to a depth of 14.9 m; they may infill a buried channel.

In the area drained by the River Brock and New Draught, there are similar wide expanses of alluvium, though the boundaries of the deposits are in places poorly defined, particularly along New Draught. Sections in the River Brock immediately north of the Lancashire College of Agriculture reveal silts and silty sands with a thin basal gravel, overlying till. Around Myerscough House, ponds have been dug through the alluvium into the underlying till, indicating that the alluvium is not much more than a thin veneer hereabouts. However, around New Draught Bridge [479 401], boreholes show the alluvium to be at least 8 m thick and to comprise sandy clay, sand and laminated clay.

West of St Michael's, the alluvium is again restricted in width because the River Wye is bounded by rising till surfaces to the north and south. The river is tidal throughout this tract and, from about half a kilometre downstram of Cartford Bridge, saltings undergo regular flooding. Exposures are poor, but in the river bank [4021 4133] south of Out Rawcliffe 2 m of grey-brown sandy silt are visible.

In the upper part of the Brock valley thin but extensive spreads of alluvial silt and sand occupy broad hollows [558 454] north-east of Tootle Hall and [579 445] east of Higher Brock Mill. These recent deposits may conceal underlying glaciolacustrine deposits, but neither the lithology nor the geometry of these sediments has been ascertained.

Ribble and Hodder valleys

Alluvium forms extensive spreads in the Ribble valley, rising to about 5 m above the normal river level. The surface of the alluvium is uneven in places and commonly shows evidence of former river courses. The deposits are graded, with a 1 to 2 m basal layer of cobble and boulder gravel and up to 4 m of fine silt or clay above. Reddish brown till fragments are incorporated locally in the gravel [e.g. at 6149 3332].

Expanses of alluvium up to 4 m thick occur in the Hodder valley upstream of Whitewell, and more sinuous tracts are present in Langden Brook and adjoining the River Loud. Sands and silts with subordinate gravels are the main sediments. On many of the wider alluvial flats there are abandoned channels containing peaty clays, for example in the Langden valley [6505 5040], in the Dunsop valley [6595 5040] and in the Hodder valley [6638 5012].

Lowland till plain

Away from the main rivers, thin patches of alluvium fill hollows on poorly drained parts of the till plain. The deposits are commonly associated with peat and, as noted earlier, are particularly prevalent where melt-out of buried ice has resulted in relief inversion.

Several larger depressions, floored by silt, clay and peat, mark the sites of ephemeral lakes. One is located [488 468] just west of Cabus village and another [465 390] lies to the north-east of Pad End Farm. Augering at the first location proved grey-brown silty clay with a thin interbedded peat, while a temporary roadside excavation [4862 4597] showed 0.4 m of smooth grey clay.

Peat

Peat formation was initiated at the end of the Devensian cold stage and has continued in different parts of the district to the present day. The earliest peat deposits formed in kettleholes on the till surface and date from before 10 000 years BP. More extensive spreads of peat began to accumulate after the mid-Flandrian marine transgression and these "mosses" cover large areas in the Pilling Water and lower Wyre catchments. At higher levels (around 100 m above OD) there are *remanié* peat deposits, as near Blackmoss House [605 399], and hill peat, in part coeval with the lowland mosses, blankets the high moorlands in the north-east of the district.

Details

Pilling to St Michael's on Wyre

Although peat cutting, coupled with land reclamation and improved drainage, has resulted in the destruction of large areas of lowland peatland, there remain extensive deposits in the mosses of Cockerham, Winmarleigh, Pilling and Rawcliffe. Of these, only Winmarleigh Moss [446 483] remains largely untouched by modern farming practice. In areas where the peat has been removed by cutting, often the only evidence for its former presence are vestiges of peat preserved beneath lanes and tracks. A good example is the stretch of lane [4320 4693 to 4490 4623] near Cogie Hill Farm, which runs for two kilometres on an elevated causeway of peat, with the underlying older estuarine deposits now forming the surface on either side.

The more northerly mosses (Cockerham, Winmarleigh and Pilling) are composed of sedge and sphagnum peat, with some woody remains at the base (Oldfield and Statham, 1965). They contain bog-oaks which are periodically uncovered as reclamation proceeds. The peat is up to 3 m thick on Winmarleigh Moss but thins against till to the south and east. On Pilling Moss the peat is generally over a metre thick, and in places exceeds 2 m, for example close to the western end of Hornby's Lane, where human activity has been restricted. Where the peat overlies estuarine deposits, it generally forms flat spreads, but on till the surface is undulatory as the peat reproduces irregularities in the surface beneath.

South of Island Farm [4590 4563] peat-filled channels separated by mounds of till probably represent a suffocated drainage network that at one time must have fed into a much extended Morecambe Bay. Thick peat is also present in a trough between the Eagland Hill Ridge and the Skitham–Trashy Hill Ridge. Here, at least 1.5 m of peat is known to be present from sections in ditches, while a borehole (SD 44 SW/5) to the north of Eskham House records 4.1 m of peat overlying till.

Rawcliffe Moss forms a topographically distinct feature quite separate from Pilling Moss. Its surface rises from below 8 m OD at the south-west corner to over 15 m in the centre. Sections in peat of up to 2 m can be seen along the western cut edge [e.g. 4277 4283]. *Remanié* peat deposits [4204 4258] to the east of Stone Check Hall are all that remain of the more continuous mosslands described from this area by De Rance (1877).

Radiocarbon analyses (Barnes, 1975) show Rawcliffe Moss to date from 5960 ± 100 BP, while a date of 3600 ± 100 BP is given for the start of peat formation near Lousanna on Pilling Moss. Samples from a borehole [4334 4358] near Skitham gave much older dates (greater than 10 000 BP), suggesting early postglacial accumulation of peat, possibly in a kettlehole. The later history of development of the mosslands, including the evidence for marine transgressions and the increasing effects of human activity, are also discussed by Barnes (1975).

Nateby – Wesham – Longridge

Isolated peat-filled depressions are particularly numerous to the east of Rawcliffe Moss and around Nateby. These occurrences, together with others in a belt extending eastwards from Esprick [410 364]

through Barton to just north of Longridge, are probably mostly kettleholes formed when buried ice melted. The main occurrences are described below.

Around Nateby [465 445] peat occurs in steep-sided hollows, the depths of which are not known. The peat probably forms only a thin cover to an infilling of clays, silts and sands. Between Nateby and around Manor House Farm [458 440] many other shallow depressions in the till surface are floored by peat.

Surrounding Wag Hill there is a large expanse of peat bounded by rising ground on its northern side and abutting river alluvium on its southern edge. Several exposures and auger probings indicate that it overlies grey silts and clays. The proximity of the River Wyre suggests that it may have accumulated in a former watercourse of that river.

An extensive spread of peat is present [504 443] in a broad hollow between Dimples and the Lancaster Canal. The hollow opens out to the south into the broad floodplain of the River Calder. The thickness of the peat is not proven.

Between Medlar [417 357], Elswick Leys [422 377] and Kellet's Bridge [466 347] several hollows and elongate depressions contain peat, probably between 1 and 3 m thick.

A group of depressions arranged on either side of the Lancaster Canal between Bilsbarrow and Barton are mostly floored by alluvium, but a number contain thin peat.

A series of peat-filled depressions, probably aligned along a former glacial drainage channel, extends from south of Middleton Hall, through Little Westfield, to Cross [550 372 to 535 373]. Some of the depressions are up to 250 m in diameter, with prominent rims of till and, in plan, resemble ice-disintegration features described by Winters (1961).

At the south-western end of the Vale of Chipping the remnants of a peat moss are preserved on the low-lying plateau [605 399] near Blackmoss House. The peat appears to have accumulated in a small basin that drained southwards through a series of anastomosing channels towards the River Loud. The deposit has been cut extensively, and being generally less than a metre thick, is not shown on the 1:50 000 map.

Bowland Fells

An extensive blanket of hill peat covers much of the high moorland in the north-east of the district. Generally only 1 to 3 m thick where it lies on moorland slopes, it reaches more substantial thicknesses in bogs and where it fills hollows. The peat blanket is being actively eroded, most obviously on the top of Hawthornthwaite Fell [580 515], but also on the summits of Fair Snape Fell [592 468], Saddle Fell [612 472] and Totridge [635 487]. Here, the once-continuous peat blanket now consists of a series of peat hags highly dissected by drainage gullies and larger peat-free areas. The causes of peat erosion are complex and manifold, and are discussed in detail by Tallis (1985).

Calcareous tufa

There are two small patches of calcareous tufa on the right bank of the River Hodder in Whitewell gorge. Both were deposited by streams issuing from nearby outcrops of Clitheroe Limestone. The northernmost deposit [6545 4655] forms a cliff of fine stalactites at the river's edge; the other [6533 4635], though less spectacular, also underlies a waterfall.

Calcareous tufa is also presently accumulating in a slipped area of till in a gully [6565 4523] north of Ing Wood, and around a spring at the back of the Third Terrace [6770 3651] of the River Ribble near Hey Hurst.

Cave deposits

Caves [6554 4678], known as Fairy Holes, with deep floor coverings of silty clay occur in the Clitheroe Limestone in the gorge of the River Hodder below Whitewell. No excavations are known and therefore no history has been established.

Blown sand

Mounds of wind-blown sand between 1.5 and 2 m high occur close to the foreshore in the north-west of the district. The main deposits can be found at Marsh Houses [4570 5143], and between Sand Villa [4365 5055] and 300 m west of Mill House [4310 4995].

Landslips

Landslips are common in the eastern half of the district, many occurring on steep rock slopes in the Bowland Fells and others involving drift deposits. Rock-cored slides are prevalent in the Langden, Grizedale and Calder valleys, and more especially on the steep escarpment between Parlick and Mellor Knoll. Slides involving superficial deposits (mainly till and head) are conspicuous in the valley systems of the rivers Hodder and Ribble, and are a potential hazard wherever rivers are undercutting relatively unconsolidated deposits.

The majority of the larger slides are rotational in type, with failure apparently having occurred along a concave surface at the back of the slip. In slides involving the Pendle Grit, bedding-plane slip along mudstone partings appears to have contributed to failure, particularly where the bedding was inclined in the direction of the ground slope.

The larger landslips, such as the one beneath Wolf Fell, are clearly of some antiquity and were probably initiated in late-Devensian times when the near-surface rock layers would probably have been water-saturated and weakened by periglacial freeze–thaw processes. Slips in glacial deposits range widely in age and are still liable to occur when slopes are either undercut by river and stream erosion or destabilised by human activities such as road construction.

Details

Langden, Dunsop and Hodder valleys

Landslips are numerous in the valley of Langden Brook and its tributaries, where they are underlain by sandstones of the Pendle Grit Formation. Mudstones within the sequence may have contributed locally to slipping; head deposits are also involved.

A group of much larger slips scars the south-eastern flank of the Bowland Fells between Mellor Knoll and Parlick. The largest, over 1 km² in extent, affects the whole of the south-east face of Wolf Fell and involves failure of Pendle Grit. Instability here has been enhanced by groundwater moving through sandstones at the the base of the Pendle Grit and lubricating the surface of the underlying Bowland Shales. Large slips [645 488] to the north-east of Wolf Fell and on the nearby Mellor Knoll mainly affect the alternating mudstone–limestone sequences of the Bowland and Worston Shale groups.

Slips in till occur along both sides of the Hodder valley and are particularly prevalent in the deeply incised streams draining south-eastwards from Totridge. Some of the slips, such as those at Hodder Bank [654 485] and near Hareden [645 505], are associated with weak, laminated clays. A slip on the right bank of the Hodder [6442 4347] is presently threatening the adjacent public road.

Ribble valley

In river cliffs along the Ribble valley there are slip-scars in till, some up to 50 m high and 200 m across. Most are active or have been active in the recent past because the river undercuts the valley sides on the outer banks of meanders. Water seepage from sand layers within the till and from the till/rockhead junction has probably helped initiate these slips. Adjoining the active slips are areas of scarred, hummocky ground believed to represent zones where minor movements or hill creep may have taken place; the northern end of Lower Madgell Bank [655 346] is an example.

Active slips are also present in the lower reaches of the more deeply dissected tributary valleys, notably in Old Park Wood [653 337] and along Duddel Brook and other, unnamed right bank tributaries. In a larger area of slipped ground [614 355] near Meadow Head, several distinct phases of movement can be distinguished.

The only major landslip involving solid strata, hereabouts, has occurred on the east bank of Duddel Brook [659 375] where siltstones and flaggy sandstones striking parallel to the river have suffered rotational failure.

Grizedale, Calder and Brock valleys

On the north side of Grizedale there are substantial slips in Pendle Grit in a place where the local dip is 20°, directed towards the valley. Otherwise slips in these valleys are confined to steep till slopes, such as occur in the River Brock at Brock Cottage Farm [549 435]. On the opposite bank [548 432] of the same river a slip is presently being eroded by the river and is intermittently active.

Made ground

Small areas of made ground are associated with former sand and gravel workings in the Wyre valley [516 512] and at Bradley's Sand Pit [511 340], where a landfill project is in operation. There are also restricted deposits of made ground alongside the M6 motorway [e.g. 542 350] where the natural land surface has been built up using glacial drift excavated from cuttings. On Longridge Fell, former limestone quarries [676 419] have been almost completely infilled with quarry and domestic waste, and a number of small sandstone quarries partially backfilled.   DMcB

SIX

Structure

The Carboniferous rocks in the eastern part of the district were folded and faulted most intensively during the Hercynian (or Variscan) orogeny at the end of the Carboniferous Period, to produce a group of structures that form part of the Ribblesdale Fold Belt (Figure 20). The existence of this system of folds was first recognised by Phillips (1836, p.106). Some earth movements also took place in this region at intervals through Dinantian and early Namurian times (Arthurton, 1984; Gawthorpe, 1987).

The Triassic rocks in the western part of the district are mostly concealed beneath Quaternary deposits and their structure is, consequently, largely conjectural. Nevertheless, it is clear from the few deep boreholes that have been drilled, supplemented by seismic reflection data*, that the Triassic rocks are mostly unaffected by the type of folding and faulting prevalent in the Carboniferous tract to the east. Instead, the structural picture is dominated by north-north-

east-trending syndepositional faults, notably the Woodsfold Fault, and by a major downfold, the Kirkham Syncline, at the south-west margin of the district. A poorly defined structure of similar trend, named the Weeton Anticline by Wilson and Evans (1990), lies at the western margin of the district (Figure 21). Wilson and Evans point out that this general north-north-easterly structural trend is subparallel to that affecting Carboniferous rocks over a wider area of Lancashire and relates also to the trend of late Silurian rocks in southern Furness and the faults that bound the graben-like basins in the Permo-Triassic of the Irish Sea (Jackson et. al., 1987). They therefore suggest that many of these structures lie on 'inherited end-Silurian lines which were intermittently active at least until Jurassic times'.

* These data are held by BGS on a 'commercial in confidence' basis and cannot yet be published.

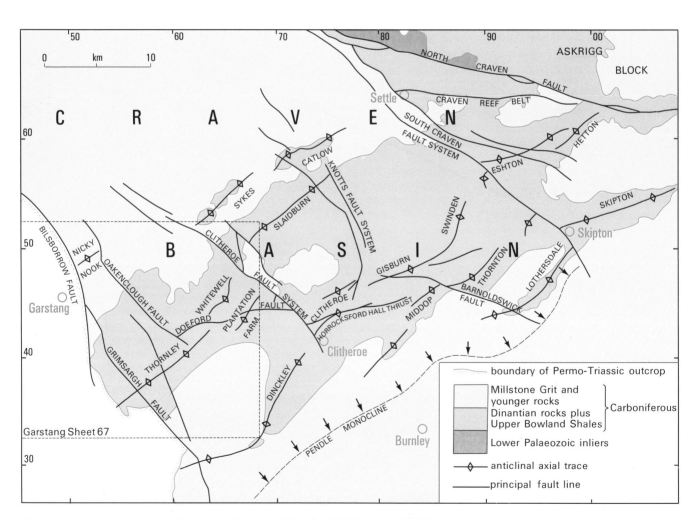

Figure 20   Location of the Garstang district within the Ribblesdale Fold Belt.

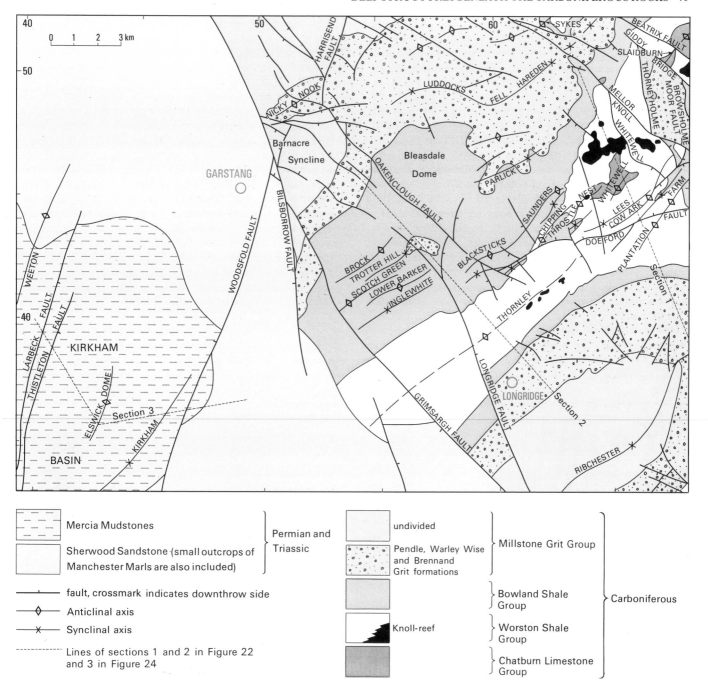

Figure 21 Main structural features of the district (for sections 1 and 2, see Figure 22; for section 3, see Figure 24).

DEEP STRUCTURES BENEATH THE CARBONIFEROUS ROCKS

Although there is no direct evidence as to the nature of the pre-Carboniferous basement rocks, they are assumed, in the absence of contrary gravity and aeromagnetic evidence, to be similar to those of the nearest exposures at the south-western edge of the Askrigg Block in the Settle district (Arthurton and et al., 1988). The Craven ('Bowland') Basin is thought by Gawthorpe (1987) to have formed initially by the subsidence of a large south-eastward tilting block bounded on the north and north-east sides by the Craven Faults, and on the south-east side by a line later occupied by the Pendle Monocline, and probably overlying a major basement fault downthrowing at that time to the north-west (see, for example, Lee, 1988b, fig 8.3). Some indication that the basement has been broken up into smaller fault-bounded tilt blocks has been provided by interpretations of regional gravity anomalies, both in adjacent parts of the Craven Basin (Gawthorpe, 1987; Lee, 1988b) and in the present district

(Lee, 1988a; Kimbell, pp.89–91 of this memoir). These intrabasinal blocks are mostly tilted north-westwards, bounded on their south-east margins by faults antithetic to the main Pendle Monocline fault and on their north-east and south-west sides by transfer faults. Interpretation of seismic reflection data acquired provides some support for the existence of the basement blocks postulated from the gravity modelling outlined above.

In the district, the broad gravity high ('D' in Figure 26) that underlies much of the central part of the Carboniferous outcrop is now shown, from the interpretation of a very sparse coverage of seismic reflection data, to be characterised by relatively shallow basement at about 2000 to 2300 m below sea level (Figure 22); it may be even shallower in places. To the south, beneath the Ribchester Syncline, this same top basement horizon lies at depths between 3900 and 4900 m. A major concealed, syndepositional fault, here named the Thornley Fault, marked by a steep gravity gradient, is therefore inferred to lie between these two areas, to account for the large difference in basement level (Figure 22, section 2). There are also a number of lesser faults which affect the basement and the overlying early Carboniferous strata but which do not appear to affect the surface outcrop. Some of these faults probably influenced the position of anticlines formed later, during the Hercynian compressive phase.

The Doeford Fault (p.86) and an associated belt of country characterised by intense folding and cleavage probably overlies the concealed Thornley Fault (Figure 22) which appears to trend generally north-east. The fault plane dips to the south-east, i.e. facing, or slightly oblique to the direction from which the main Hercynian compression was probably directed. It is envisaged that these movements compressed (or transpressed) the more argillaceous and ductile Carboniferous sediments against tougher Lower Palaeozoic rocks at the faulted block margin, producing the observed folding and cleavage, reversing the movement of the synsedimentary fault and propagating the Doeford Fault.

INTRA-CARBONIFEROUS MOVEMENTS

Direct evidence for intra-Carboniferous movements affecting the district is sparse, but there is good reason to suppose that tectonic events which have affected other parts of the Ribblesdale Fold Belt would have left some imprint here. The best of such evidence is provided by sedimentological studies of the rocks themselves (Gawthorpe, 1986), which reveal, for example, the presence of conglomeratic carbonate rocks that indicate a sudden increase in relief at the basin margin due to movement of marginal faults. Another example is the presence of intraformational slide and slump structures. These could have been caused either by earthquakes or by an increase in the slope of the sea floor on which the sediments were being deposited, or simply by rapid deposition leading to instability in the soft-sediment pile. Tilting of the sea floor could have occurred in response either to penecontemporaneous folding during a phase of compression or transpression, or to the movement of fault-bounded crustal blocks during a phase of crustal extension. Tectonic movement is also indicated by unconformities and non-

sequences in the rock succession. The most widespread of such breaks in the Craven Basin occurs at the basal contact of the Hodder Mudstone and the Clitheroe Limestone formations (Riley, 1990b). Studies by Riley of fossil assemblages above and below this break show that it is intra-Chadian in age, which implies a tectonic episode of this age. Other breaks occur locally at higher levels.

Most of the features mentioned above, which probably reflect tectonic acivity within and marginal to the Craven Basin, could also have been caused by regressive eustatic sea-level changes (Ramsbottom, 1973); some element of this effect is not ruled out.

Within the district, good evidence for tectonic movement in mid-Chadian times is represented by the unconformity, mentioned above, between the knoll-reefs of the Clitheroe Limestone Formation and the overlying beds of the Hodder Mudstone Formation, including conglomerates and boulder beds in the Limekiln Wood Limestone (pp.19–25). The boulder beds were mostly derived from the adjacent and underlying knoll-reefs, when the latter were at or above wave base during an interval of relatively low sea level. The lowering may have been due to either eustatic or tectonic causes, as discussed above, but the evidence from the Craven Basin as a whole (Arthurton, 1984; Gawthorpe, 1987) tends to favour a tectonic cause and this is given further support by the gravity modelling of Lee (1988a and 1988b). In particular, one of Lee's model profiles indicates that the Thornley knoll-reef belt overlies the margin of a (pre-Carboniferous) basement horst which probably moved soon after deposition of the knoll-reefs. Conglomeratic Limekiln Wood Limestone is also well exposed on the left bank of the River Hodder [6520 4450], south-west of Ing Wood, and is probably also associated with the south-east margin of this horst.

STRUCTURE OF THE CARBONIFEROUS ROCKS AT OUTCROP

Folds

The Carboniferous rocks cropping out in the central and eastern parts of the district have been folded into a series of anticlines and synclines with a predominant north-east axial trend, and have been subjected to movements along faults with a general north-west trend (Figure 21). The outcrop forms the south-western part of the Ribblesdale Fold Belt (Figure 22). The following account gives the evidence for the existence and general form of each major fold.

The **Nicky Nook Anticline**, in the north-west part of the district, is marked by an upstanding promontory (Nicky Nook) of Pendle Grit whose east-north-east-trending crest probably coincides with that of the anticline. The structure is somewhat asymmetrical, verging to the south, with sub-parallel normal faults on its southern limb (Figure 21). The fold plunges to the west-south-west at about 12°. Its continuation to the east-north-east across the Harrisend Fault is indicated by south-easterly dipping beds in the Grizedale Brook, but inadequate exposure precludes further delineation of the axial trace. The isolated position of this anticline and the steepness of its limbs, in contrast to the broad Barnacre Syncline immediately to the south, remain problematical.

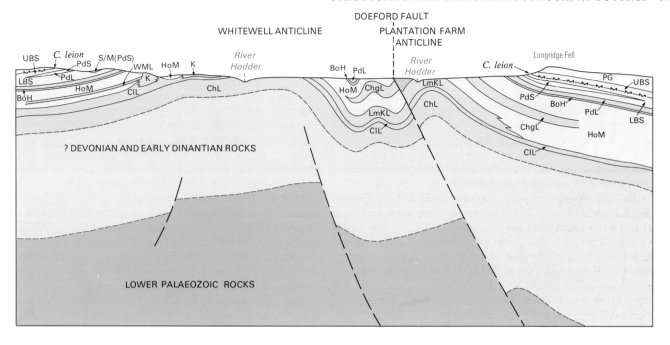

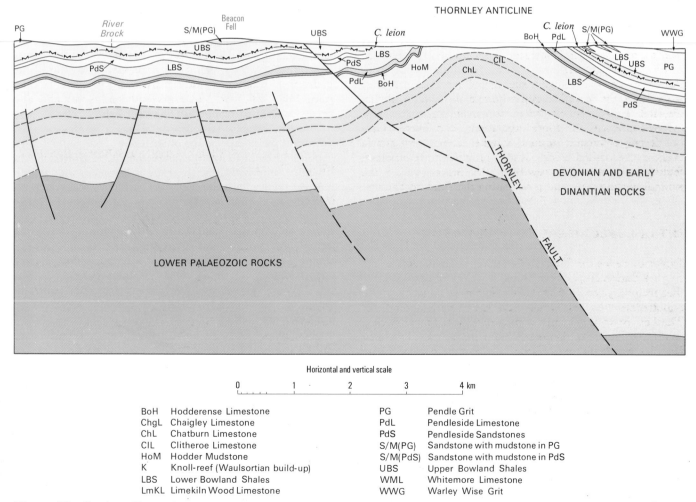

Horizontal and vertical scale

0   1   2   3   4 km

BoH	Hodderense Limestone	PG	Pendle Grit
ChgL	Chaigley Limestone	PdL	Pendleside Limestone
ChL	Chatburn Limestone	PdS	Pendleside Sandstones
CIL	Clitheroe Limestone	S/M(PG)	Sandstone with mudstone in PG
HoM	Hodder Mudstone	S/M(PdS)	Sandstone with mudstone in PdS
K	Knoll-reef (Waulsortian build-up)	UBS	Upper Bowland Shales
LBS	Lower Bowland Shales	WML	Whitemore Limestone
LmKL	Limekiln Wood Limestone	WWG	Warley Wise Grit

Figure 22   Sections illustrating the geological structure.

The **Barnacre Syncline** has an indefinite semicircular form in plan that precluded the plotting of any linear axial trace. The axial part of the fold is marked by the presence of the youngest (Arnsbergian) formations in the area, particularly the fossiliferous Caton Shales, together with the inferred outcrops of the overlying Heversham House and Crossdale Shale formations. The outcrops of these youngest strata are also bounded by faults, which have produced grabens within the overall synclinal structure.

The **Luddocks Fell–Hareden Syncline** is a broad symmetrical fold whose axial trace follows a sinuous but general north-easterly course for 7 km across the central part of the Bowland Fells Pendle Grit outcrop. It is defined in the west around Luddocks Fell [573 490] where dips measured on both limbs, in the River Calder and its tributary the East Grain, average 16°. Eastwards from here, the axial trace position is constrained by dip evidence in the deeply incised valleys of the Bleadale Water and Hareden Brook on either side of Hareden Fell [619 491].

The presence of a poorly defined and largely drift-covered structure here, named the **Bleasdale Dome**, is indicated by the general disposition of the Upper Bowland Shales and Pendle Grit Formation outcrops around Bleasdale [57 45]. The inferred outcrop of the Pendleside Sandstones occupies the central part of the structure, where there are few exposures. Dips in these are low (3° to 6°), with the exception of a 25° dip near a probable fault line. The north-west-trending **Oakenclough Fault** is inferred to cross the structure on its south-west flank.

The **Parlick Syncline** to the east is a broad gentle structure revealed by its effect on slack features on the Pendle Grit escarpment at the indented margin of the outcrop on the south side of the Bowland Fells; it can be traced for about 3 km.

A series of three anticlines and two synclines, with general east-north-east axial trends, is present in the Bowland Shales outcrop to the west and south-west of Beacon Fell. These comprise, from north-west to south-east: The Brock Anticline, the Trotter Hill Syncline, the Scotch Green Anticline, the Lower Barker Anticline and the Inglewhite Syncline. The axial part of another unnamed anticline, inferred from exposures near Cross House Claughton [5289 4291], is also present in this tract of country. Almost certainly, other folds of comparable style to these lie undetected beneath the drift in this part of the Bowland Shales outcrop.

The axial trace of the **Brock Anticline** appears to be immediately south-east of the course of the River Brock from [5317 4130] near Walmsley Bridge to Brock Bottom [5469 4225]. The crest of the fold is exposed in the river gorge at the former locality in tough, thinly bedded, calcareous siltstones. Minor faults with steep hades and directional trends between N026° and N060° are associated with the crest of the anticline here. Other more substantial faults with trends between N116° and N115° are exposed upstream [e.g. 5344 4150], affecting beds on the north-north-west limit of the fold, where dips average about 14°. The south-south-east limit is largely unexposed.

The next identified fold to the south-east is the **Trotter Hill Syncline**. This is most clearly defined where it affects the base of the Pendle Grit on the west side of Beacon Fell [562 424]. The south-south-east limb here appears much steeper than the other limb. The same fold helps to explain the presence of an outcrop [5455 4151] near Trotter Hill of what is inferred to be the Hind Sandstone.

The presence of the **Scotch Green Anticline** is revealed firstly by exposures of tough calcareous siltstones in the highest part of the Lower Bowland Shales sequence, which dip at 29° to 45° on the northern limb of the structure between Lower House and Winn House [5364 4078 to 5403 4097]. Secondly, the south-westerly plunging crest of the structure forms a promontory feature [534 402] east of Spaddock Hall, the plunge being about 6°.

The **Lower Barker Anticline** appears as a subsidiary fold on the south-east limb of the Scotch Green Anticline. The axial trace is interpolated towards the north-east across the largely drift-covered outcrop of the Upper Bowland Shales on the evidence of dips in stream exposures around Lower Barker [553 409] and Fell Side [570 417]. Dips of 50° on the south-east limb at the former locality indicate a marked asymmetry verging towards the south-east.

The **Inglewhite Syncline** is the most southerly structure detected in this group of folds. It appears to have a broadly symmetrical form but only beds on the south limb in the Sparting Brook [near 548 392] and around the north-easterly termination [570 411], north-east of Ashes, are exposed. Dips of 30° to 35° are common in both these areas.

All the folds in the group described above appear to terminate to the north-east near the Longridge and Beacon Fell faults. However, there is little exposure between these faults and the subparallel Oakenclough Fault lying to the east of Beacon Fell, and the structures are consequently not known in detail. It is assumed that the Blacksticks Anticline and the complementary syncline to the south-east persist with a south-westerly plunge across this fault line.

The **Blacksticks Anticline** is in fact a symmetrical anticlinorium with dips on the outer limbs of 52° to 64° in the Pendleside Sandstones, the main member involved in the structure. Internally, the structure is complex, with at least four subsidiary anticlines (A1–A4 in Figure 23), which are seen in discontinuous exposures in the gorge at Hough Clough [5938 4256 to 5953 4244], east of Blacksticks. Subsidiary anticlines A1 and A3 have a chevron form with interlimb angles of 90° and 70° respectively; A4 appears to be almost isoclinal. In contrast, A2 is a broad, saddle-shaped, open fold with a narrow, tight syncline separating it from A3. Calcite veining and brecciation associated with the hinges of anticlines A1, A3 and A4 probably represent accommodation faulting subparallel to the axial planes of these non-cylindrical folds. Bedding plane shear is also evidenced by calcite veining in places. These subsidiary folds tend to have axial trends that subtend angles of 26° to 28° to that of the major structure. Moseley (1962) and Arthurton (1984) have recorded similar oblique axial trends in the Sykes and Catlow anticlines (Figure 20). Towards the north-east, the Blacksticks Anticline may be represented by the Saunders Anticline in the valley of the Chipping Brook. The structure hereabouts is poorly defined, however, owing to the absence of exposure around the fold hinge, which is marked by a conjectural outcrop of the Pendleside Limestone [616 437]. The hinge length of the combined folds is nearly 6 km.

The **Chipping Anticline** to the south is revealed by exposures in the Chipping Brook in Chipping village and in the

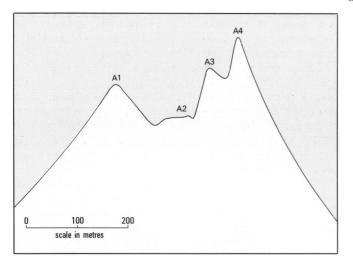

Figure 23 Schematic profile showing the general form of the Blacksticks Anticline.

stream north-west of Radcliffe Hall [617 426]; both sections are in the Pendleside Limestone outcrop, which marks the core of the otherwise poorly defined structure. The fold probably verges to the north-west.

The **Thornley Anticline** is the largest structure in the district but also the least known owing to the extensive drift cover. Its presence is indicated by the disposition of the largely conjectural outcrops of the Lower and Upper Bowland Shales, which in turn define the Hodder Mudstone outcrop at the core of the structure (Figure 22, Section 2). The hinge length is at least 11 km; the fold presumably continues beneath the unconformable Permo-Triassic cover to the south-west. The position of the axial hinge line shown in Figure 21 is conjectural because there is no good evidence. The oldest strata known to crop out are those of the isolated group of knoll-reefs of Chadian age in the basal part of the Hodder Mudstone between Knott Farm [613 402] and Leach House Farm [628 414] . The knolls are aligned parallel to the axial trace and probably crop out through the drift cover because of their upstanding character and resistant lithology. However, coeval or somewhat older strata may be present nearby to the north-west beneath the drift. To the south-west the only other significant exposure in the Hodder Mudstone is in the Westfield Brook [5514 3796], where tight folds in a limestone/mudstone turbidite sequence of Arundian age verge towards the west and plunge steeply to the south. These north–south-trending minor folds are at variance with the general north-easterly fold trend, possibly due to stresses associated with the nearby Grimsargh Fault. To the north-east, the axial trace of the Thornley Anticline has been inferred to continue some distance beyond the knoll-reef belt and may plunge and disappear in that direction. There is a possibility, however, that the outcrop of the Limekiln Wood Limestone at High Head Wood [635 427] represents the north-easterly plunging core of the structure, truncated on its northern side by the Doeford Fault.

The **Sykes Anticline** lies mainly in the district to the north; only the extreme south-western end of the structure is present near the northern margin of the present district [625

518]. The anticline plunges to the south-west and apparently terminates against the Mellor Knoll Fault (Figure 21). A relatively weak anticlinal structure, dextrally offset by this fault, can be traced in the Pendle Grit outcrop to the west.

The Sykes Anticline has been described in detail by Moseley (1962), who emphasised the disharmonic relationship between the upper level of the fold, as it affects the Pendle Grit, and the lower level involving the Dinantian limestones, with a structurally incompetent sandwich of Bowland Shales between. Moseley illustrated this disharmonic relationship with a series of cross-sections which show a strong asymmetry and north-westward vergence at the lower level. The vergence is also shown by the way the axial trace is offset north-westwards from the Hetton Beck Limestone outcrop in the core of the structure (Figure 21). Further details are given by Hughes (1986).

The **Whitewell Anticline** lies to the north-east of the Thornley Anticline in an *en échelon* relationship. It has a highly irregular periclinal form, as shown by the indented outcrop of the Chatburn Limestone, which forms the core of the structure (Figure 21). Within the main structure, three subordinate *en échelon* anticlines are recognised. These are centred just west of Greystoneley Brook [644 449], constituting the north-easterly plunging termination of the Throstles Nest Anticline, in the gorge of the River Hodder below Whitewell [655 466] and to the east of the Hodder [658 455]. The best-defined structure is the last of these, named the New Ground Anticline by Fletcher (1990).

To the south-east of the Whitewell Anticline there is a series of major folds, including the **Lees Syncline**, the **Cow Ark Anticline** and the **Plantation Farm Anticline**. The strata involved in these structures belong mainly to the more ductile argillaceous parts of the Dinantian sequence. The folds are correspondingly tight and narrow, with steep dips, and cleavage is developed locally, especially on the north-west limbs of the Cow Ark and Plantation Farm anticlines [e.g. 6471 4343 (Plate 14); 6691 4397; 6574 4317 to 6608 4310]. The intensity of the folding and the presence of cleavage (Plate 15) in this tract of country cannot be fully explained by the argillaceous lithologies involved, however (see below). The Lees Syncline has vertical and slightly overturned beds on its south-east limb, where exposed in the River Hodder [6465 4349 to 6471 4334]; the north-west limb is somewhat less steep with dips of 47° to 65° near [658 443] Middle Lees. The Cow Ark Anticline is indicated mainly by the flanking *en échelon* Lees and Mill Brook synclines (Fletcher, 1990), rather than by the disposition of geological boundaries around the core of the structure.

The axial traces of subordinate folds within these major structures appear to be mainly subparallel to the main axial trends. An exception is on the southern flank of the Thornley Anticline, in a 2.5 km tract to the south-west and north-east of Thornley Hall, where the trends of numerous minor folds exposed in streams [e.g. 6419 4159 to 6162 4085] are mainly between N177° and N189°. Such folds would subtend angles of about 34° to 46° to the inferred axial trace of the poorly defined Thornley Anticline and would be similarly oblique to the general trend of the Ribblesdale folds.

To the south-east, between the Thornley Anticline and the Dinckley Anticline in the Clitheroe district, lies a broad, well-defined fold named the **Ribchester Syncline** (Bridge,

Plate 15 Cleaved mudstones in the uppermost beds of the Hodder Mudstone Formation exposed on the left bank of the River Hodder [6474 4341] near Stakes. The cleavage here affects only a few metres of beds immediately below the Hodderense Limestone Formation (A14824).

1989). The north-eastward termination of this structure is rimmed by the well-marked escarpments of the Pendle and Warley Wise grits, and its axial area in the low ground of the Ribble valley coincides with the outcrop of the lower part of the overlying Sabden Shales (Figure 21). The syncline is asymmetrical, with dips in the sandstone outcrops on the north-west, north-east and south-east flanks in the range 10° to 31°, 10° to 28° and 38° to 50° respectively. In the Triassic Sherwood Sandstone outcrop to the south-west, a syncline east of Preston, noted by Price et al. (1963, p.80), seems to continue the axial trend of the Ribchester Syncline. This would imply a resumption of Hercynian fold movements in Triassic times or later (see p.87 and Wilson and Evans, 1990).

Faults

Faults affecting the Carboniferous outcrop are described from west to east, with evidence for their existence and form. Figures given for the amount of displacement are estimated from mapped field relations only.

The boundary between the Permo-Triassic and Carboniferous outcrops, which trends in a general north-north-west direction across the centre of district, is considered to be partly an unconformity and partly a fault, here named the **Bilsborrow Fault**. The line of this fault, like that of most of those in the district, has not been precisely fixed, but is inferred from the evidence of boreholes, such as those around Brock and Stubbins (SD 54 SW/7, 5A and 21) and around Parkhead Farm [508 454]. Although the spacing of these boreholes is too great to prove that the boundary is a fault rather than an unconformity, the former interpretation is preferred where shown on the map because the inclination of the boundary is evidently greater than that of the plane of unconformity, which is generally no greater than 5° in the district. The actual fault was probably penetrated in only one borehole, namely SD 54 SW/1 near Garstang. The driller's record showed 'red marl', assigned to the Manchester Marls, from 10.97 to 19.35 m depth, overlying 'broken bastard rock' to 21.34 m, probably the fault breccia. This overlies grey and red sandstone and grey shale, assigned to the Roeburndale Formation, which were seen to the bottom of the hole at 34.75 m.

Farther north, evidence from other boreholes, such as SD 44 NE/7 and 13 near Cabus, help to constrain the position of the fault line. The former borehole, in particular, proved Sherwood Sandstone from rock head to the bottom of the hole at 37.5 m, at a site only 250 m from small exposures [5017 4788] of Carboniferous Park Wood Sandstones to the east of the fault line. Steeply dipping, red-stained, turbiditic sandstones exposed [5092 4525] near Parkhead Farm and assigned to the Stonehead Sandstones also indicate the proximity of the fault.

The **Grimsargh Fault** extends north-westwards from the adjacent Preston district, where it clearly displaces the outcrops of the Wilpshire (or Warley Wise) Grit on the flanks and crest of the Dinckley Anticline, with a downthrow to the south-west. Its north-westward continuation to form the boundary between the Sherwood Sandstone outcrop to the south-west and the Carboniferous outcrop to the north-east is conjectural owing to the thick drift cover and sparsity of borehole data. Near Haighton Hall [577 353], however, the position of the fault is suggested by seismic data, interpreted by D W Holliday (personal communication, 1988). Farther north, the fault has caused a major displacement of the Pendle Grit outcrop near Claughton, as indicated by the records of boreholes SD 54 SW/19 and 23. The former record shows a mainly mudstone sequence where the Pendle Grit would have been expected had there been no fault; the latter record shows a sandstone sequence of similar facies to the Pendle Grit where Upper Bowland Shales might have been expected. The vertical displacement of the fault hereabouts is estimated to be about 230 m.

The **Longridge Fault** trends north-north-west beneath the western outskirts of Longridge with a large downthrow to the east. The presence of the fault is indicated from the interpretation of seismic evidence (D W Holliday, personal communication 1988; Bridge, 1988b). The Longridge Fault is thought to terminate against a north-north-east-trending fault crossing the eastern slope of Beacon Fell [574 420]. The line may continue, however, as the Beacon Fell Fault, a relatively well-defined structure displacing the outlier of Pendle Grit capping Beacon Fell; it has a south-easterly throw of about 15 m (Aitkenhead, 1990b).

The **Oakenclough Fault** also has a north-westerly trend, displacing the Pendle Grit near Oakenclough [537 472] and Fell End [547 462] and largely causing the south-westward termination of the Pendleside Sandstones outcrop in the Blacksticks Anticline, near Blacksticks [591 421]. The presence of this fault was mainly confirmed by feature mapping; there is no borehole control. Some geochemical

anomalies in the Dawshaw to Little Elmridge area may be associated with this fault (Wadge et al., 1983). North-west of Oakenclough the fault appears to splay into several subsidiary fractures, some on the south flank of the Nicky Nook Anticline and others, including the **Harrisend Fault**, traversing the anticline.

Though evidence on the western (downthrow) side of the fault is obscured by drift, the Harrisend Fault appears to be a major structure with a throw in the order of 400 to 500 m, terminating the Bowland Fells Pendle Grit outcrop over a distance of about 2 km at the foot of the western slopes of Harrisend Fell [531 503].

All the other major north-west-trending faults in the district lie in the north-east corner and are mostly branches of the important **Clitheroe Fault System** (Figures 20 and 21), the Clitheroe Fault Zone of Gawthorpe (1987). The **Mellor Knoll Fault**, in particular, is the direct continuation of the Clitheroe Fault mapped by Earp et al. (1961) in the adjacent district to the east. This fault system also includes the Whitewell, Thorneyholme, Giddy Bridge and Browsholme Moor faults. The system probably extends several kilometres to the north-west into the Lancaster district. Details of this system are described in the BGS Technical Reports for the relevant 1:10 000 geological maps (Fletcher, 1990 and 1991). The main effects are the termination of the Slaidburn Anticline at its south-west limit on the upthrow side of the Giddy Bridge Fault, and the production of a graben within the system. This structure, here named the **Birkett Fell Graben**, is bounded by the last-named fault together with the Browsholme Moor, Mellor Knoll (southeast part) and Thorneyholme faults. Paradoxically, the downthrown younger formation within the graben, being capped by the Pendle Grit, gives rise to the high moorland of Hodder Bank Fell [673 491] and Birkett Fell [672 483].

The **Giddy Bridge Fault** has the greatest amount of downthrow, apparently in the order of 1000 m, around the axial part of the Slaidburn Anticline, where the fault throws Chatburn Limestone against Pendle Grit. If, however, this anticline developed to a greater or lesser extent during Dinantian deposition (Arthurton, 1984), i.e. before the Pendle Grit was deposited, the effective downthrow would be correspondingly less. Certainly, some 1.7 km to the southeast, beyond a point where the fault bifurcates, the combined throw of the two branches is only about 110 m. This apparent diminution of the amount of downthrow may be because the Dinantian sequence is attenuated near a basement high in the Ashnott area (Clitheroe district).

On the west side of the aforementioned graben, the **Thorneyholme Fault** has a maximum estimated downthrow to the east of about 300 m. The **Mellor Knoll Fault**, to the west, has an estimated throw of 115 m near its junction with the Thorneyholme Fault [6679 4726]. The throw increases to a maximum of about 300 m north-west of Mellor Knoll, just beyond the point [6455 4950] where it is joined by the Whitewell Fault which, to the north-west of Whitewell, has a throw of about 100 m. Further to the south-east, however, this direction of throw is reversed, with a hinge zone in the vicinity of Cat Knot Plantation [6604 4615]. The Mellor Knoll Fault is exposed at two places, [6383 5012] in the Langden Brook valley and [6262 5079] south-west of Hareden.

Faults trending in directions subparallel to the major north-easterly trend of fold axes include the Cow Ark Fault and the Doeford Fault, both mainly affecting Dinantian and early Namurian strata in the eastern part of the district. The **Cow Ark Fault** extends north-eastwards from the poorly defined axial core of the Cow Ark Anticline and increases its throw, so that near its junction with the Clitheroe Fault it is throwing Pendleside Sandstones to the east against Hodder Mudstones to the west, a vertical displacement of at least 100 m.

The **Doeford Fault** extends from its intersection with a branch of the Oakenclough Fault [596 415], near Little Elmridge, to the eastern margin of the district [684 442] near Ayxa Hall and continues eastward to meet the Clitheroe Fault north-west of Waddington. At depth, the fault is probably associated with the concealed Thornley Fault (Figure 22), which marks the southern boundary of the South Fells Tilt Block (Lawrence et al., 1987, fig.3). The fault is exposed in dark calcareous mudstones on the right bank of the River Hodder at its confluence with the River Loud [6498 4310] and is represented by a 1 cm wide gouge; it is possibly one of several associated fractures. Elsewhere, the fault is largely conjectural owing to the thick and extensive drift cover. It is believed to be a high angle reverse fault with a northerly downthrow and a dextral shear component, indicated by the orientation of lamination in the mudstone adjacent to the gouge at the aforementioned exposure. Reverse or high angle thrust faulting is present along the strike to the east in a 4 km wide belt between the Clitheroe and Grindleton anticlines (Earp et al., 1961).

STRUCTURES OF THE PERMO-TRIASSIC OUTCROP

Faults and folds that affect the Permo-Triassic rocks in the western part of the district, and which are therefore post-Hercynian in age, have been detected or inferred beneath the thick and extensive drift cover. The folds show no association with the dominant north-westerly trend that characterises the Ribblesdale Fold Belt in the Carboniferous rocks to the east. Instead, they seem to have resulted mainly from penecontemporaneous rifting during Permo-Triassic times along a north-north-easterly trend controlled probably by fault lines inherited from the underlying Lower Palaeozoic basement (see below).

The broad elements of the structure have been revealed by the interpretation of a few of the numerous seismic reflection lines that have been run across the area, with some control from scattered boreholes and additional information from gravity interpretation. Interpretation of all the existing data presently held on a 'commercial in confidence' basis would undoubtedly reveal a much more detailed structural picture.

The most prominent fold, the **Kirkham Syncline**, was first described by Wilson and Evans (1990, fig. 15), who inferred its presence from the stratigraphical sequence and relatively steep dips proved in the Kirkham Borehole, and from a large gravity low (see Figure 25). These authors depicted the syncline as a broad basin-like structure, defined by the outcrop of the Mercia Mudstone/Sherwood Sandstone boundary, with its deepest part between Kirkham and

Elswick, and with a steeply dipping south-eastern limb passing through the Kirkham Borehole. This configuration has now been considerably modified to take account of some of the available seismic evidence, so that the Kirkham Syncline is more restricted, with an axial trace aligned north-east through Kirkham. The broader structure, now named the Kirkham Basin, is regarded as much as a Permo-Triassic depocentre as a structural entity, with its locally enhanced deposition structurally controlled by the bounding Woodsfold and Thistleton faults. Indeed, the centre of the basin, rather than being synclinal in form, is marked by the presence of a broad anticlinal structure here named the **Elswick Dome** (Figures 21 and 24).   NA

The section in Figure 24 has been drawn to illustrate the structure, the structural history and the control of gross sedimentation in the Kirkham Basin. The most important aspect of the section is the linked north-north-easterly fault systems which have resulted in the formation of this important Permo-Triassic extensional basin or graben. The Woodsfold Fault, located towards the east end of the section, is a major nonplanar (master) basin-bounding syndepositional fault with the hade and downthrow to the west. To the west, a complex series of generally nonplanar faults is developed, displaying easterly hade and downthrow directions and therefore antithetic to the Woodsfold Fault. These faults diverge upwards and the most westerly, the Thistleton Fault, forms the western limit of the main Permo-Triassic depocentre. Increased sediment thicknesses into the graben across the bounding faults are noted at all levels within the Permo-Triassic, implying normal syndepositional fault movements from early Permian times. Minor thickness variations within the Carboniferous (Dinantian) are also observed across the faults and this may indicate that the faults are related to deep fracture systems which have had some active control over sedimentation since at least early Carboniferous times.

Within the graben, sequences of the Sherwood Sandstone Group and Manchester Marls show roll-over ('r' in Figure 24) into the main (master) and antithetic faults. Roll-over and thickening of sequences into the upper part of the Thistleton Fault plane is particularly well developed, as expressed in the broad flexure of the Elswick Dome.

Lower sequences of the Sherwood Sandstone Group show roll-over and sediment thickening into a buried (synthetic) fault to the west of the master fault. Some of the fault strands forming the antithetic fault set diverge upwards and are associated with the collapsed crestal region of the Elswick Dome. It appears that Carboniferous sequences in the footwall blocks to the main graben-forming faults have been deeply eroded, more so than the Carboniferous strata preserved within the graben. Over the footwall block to the Thistleton Fault, the Permo-Triassic sequence thins and rests with marked angular unconformity upon apparently folded Namurian rocks. To the east of the Woodsfold Fault, a thin Namurian succession is preserved beneath the Permian. Some of the erosion of the footwall block here may be attributed to footwall uplift during faulting (e.g. Jackson and McKenzie, 1983).

Overall, the graben system exhibits a morphology and a number of other characteristics typical of negative inversion structures, as described by Harding (1985). Such structures

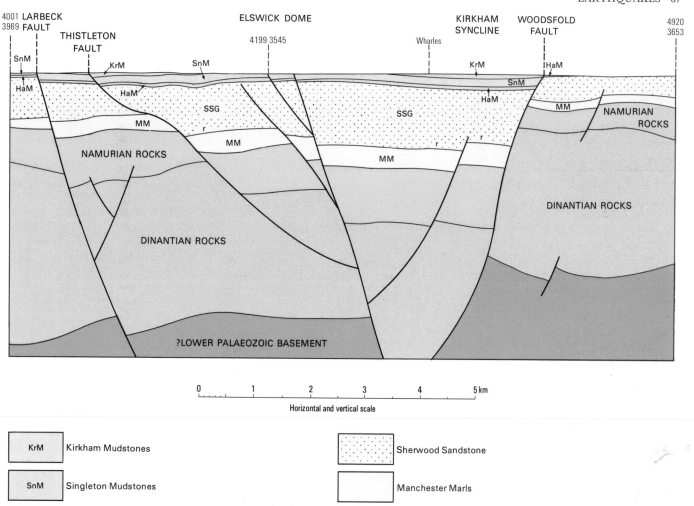

Figure 24 Section showing the geological structure beneath the Triassic outcrop along lines 1 and 2 showing in Figure 21.

are the result of divergent or transtensional movements of fault blocks. It is difficult to ascertain whether the extension here was due to dip slip only or to dip slip plus a component of strike slip, resulting from involvement of transtensional movements during the evolution of this structure. Transtensional elements are likely because the Woodsfold Fault displaces Carboniferous strata, variations in the thickness of which are observed across it. It is, therefore, likely that this fault has a long and complex movement history, affecting sedimentation intermittently through both Carboniferous and Permo-Triassic times. Chadwick (1985; 1986) has observed that major Mesozoic faults, when developing apparently as a result of extensional reactivation of earlier Variscan and older fault zones, have probably experienced a component of strike-slip displacement. This is particularly so when the extension vector is not perpendicular to the older structures.

A planar normal fault, the Larbreck Fault, lies near the north-west end of the section. It has a hade and downthrow to the east and it affects all stratigraphical levels. Some associated minor antithetic faulting affects only late Dinantian and early Namurian sequences.
DJE

EARTHQUAKES

The Garstang district, in common with many other parts of Britain, has been affected by minor earthquakes in historical times. Records have been compiled by several authors, notably Davison (1924) and Burton et al. (1984a). Records relevant to the present district are shown in Table 1. This list includes only earthquakes with epicentres in northern England. Earthquakes with more distant epicentres (e.g. that of 19 July 1984 in North Wales) may have been felt weakly in the Garstang district.

These records are based on reports of felt effects rather than instrumental data. The earthquake intensities, ranging from 3 to 5 on the MSK scale, were evidently not enough to produce any damage, although that on 20 Aug. 1835 was reported in Garstang to have 'caused many of the in-

Table 1	Date	Macroseismic epicentre (approximate)	Inferred maximum intensity in Garstang district MSK scale	Locations where felt in Garstang district
Earthquakes affecting the Garstang district (source R M W Musson, unpublished BGS records, 1990).	2 Apr. 1950	Chester	3	—
	14 Sept. 1777	Manchester	5	—
	9 Dec. 1780	Wensleydale	4	—
	11 Aug. 1786	Whitehaven	4	Garstang
	20 Aug. 1835	Lancaster	5	Garstang, Whittingham
	10 Mar. 1843	Rochdale	4	—
	17 Mar. 1843	offshore Lancashire	5	Garstang
	17 Mar. 1871	North Pennines	5	Garstang
	10 Feb. 1889	Bolton	3	Longridge
	14 Jan. 1933	Wensleydale	4	—
	30 Dec. 1944	Skipton	4	Garstang, Longridge
	9 Aug. 1970	Kirkby Stephen	4	Garstang
	30 Mar. 1972	Todmorden	3	Barton

habitants to start out of bed in alarm'. The MSK scale is considered in detail by Burton et al. (1984b) and the effects corresponding to intensities 3 to 5 may be summarised as follows:

3 — weak vibrations felt indoors by a few people
4 — moderate vibrations felt indoors by many people
5 — strong vibrations felt by many people indoors and outside, some slight damage possible.

All the earthquakes listed in Table 1 appear to have had epicentres outside the Garstang district and it is beyond the scope of this memoir to speculate on their association with known geological structures. NA

SEVEN

Geophysical investigations

The main geophysical data sets for the district have been provided by regional gravity and aeromagnetic surveys; interpretations of these provide information on the larger geological structures. An area larger than the Garstang district has been selected to provide a regional setting for interpretation. In addition to the regional surveys, more detailed local geophysical surveys, carried out to investigate oil, mineral and water resources, are reported.

PHYSICAL PROPERTIES OF ROCKS

Typical densities for the main rock types assembled from BGS data and published sources (principally Barker, 1974; Lee, 1988b and Arthurton et al., 1988) are presented in Table 2. These show that the main density contrast is between the Sherwood Sandstone and the other rock units. Within the Carboniferous sequence there are smaller variations between individual facies. The nature of the underlying basement is unknown, the nearest borehole to penetrate the base of the Carboniferous being at Holme Chapel [861 288], near Rochdale, and the nearest outcrops being those adjacent to the Craven faults in the Settle district.

Both the Permo-Triassic and Carboniferous sediments appear to be too weakly magnetised to produce detectable magnetic anomalies.

Table 2 Summary of density values for the main rock types.

Age and lithology	Density (Mg/m^3)	Number of samples
Triassic		
Mercia Mudstone	2.60	2
Sherwood Sandstone	2.26	190
Permian		
Manchester Marls	2.40 – 2.50	unknown
Carboniferous		
Namurian sandstone	2.53	12
Dinantian mudstones	2.45 – 2.60	3
Dinantian limestones	2.68 – 2.69	7
Lower Palaeozoic		
Silurian	2.70 – 2.75	
Ordovician	2.67 – 2.78	

REGIONAL GRAVITY

Bouguer gravity anomaly data are available from BGS regional surveys and from more detailed surveys released to BGS by commercial companies. The gravity coverage is good for the whole district; an average of almost 1.3 stations per km^2, with maximum coverage (2.5 stations per km^2) in the Bowland Fells area. A Bouguer correction density of $2.55\ Mg/m^3$, used to reduce the gravity data to sea-level datum, is an average value for the Namurian sandstones and Dinantian mudstones which produce most of the local topography.

The resulting Bouguer anomaly map provides considerable information on the subsurface geological structure of the region and, in order to emphasise the more local gravity anomalies, regional fields have been generated by least-squares surface fitting and subtracted from a grid of the original data (gridded at an interval of 0.5 km) to leave residual anomalies. A range of surfaces was generated, using 2nd to 10th order non-orthogonal polynomials (the higher the order, the shorter are the anomaly wavelengths remaining in the residual anomaly map). The residual map shown in Figure 25 was created using a 4th order polynomial, which was found to be the most satisfactory. The main gravity features (A to K) are annotated on this map for identification in the following discussion (see also Figure 22).

The most pronounced anomaly on the gravity map is the low centred on the Kirkham Basin (A). Evidence from boreholes, seismic reflection traverses and previous gravity modelling (Barker 1974, Lee 1988a) indicate rapid thickening of Permo-Triassic sediments in a fault-bounded basin (p.56). At the centre of the anomaly the Permo-Triassic sediments could reach thicknesses of over 2500 m. The anomaly is part of an elongated low extending from near the south edge of the map [38 25] north-north-east to the Permo-Triassic/Carboniferous boundary north of Garstang [52 47], possibly continuing northwards into the low at A' but with an offset along the line of the Bilsborrow Fault. The steep gradients which mark the eastern edge of this trough indicate a major north-north-easterly fault (Woodsfold Fault), with a large downthrow to the west. Seismic evidence and gravity modelling indicate vertical displacements of 500 to 1000 m in the section between Garstang and Kirkham (Figure 24). The gravity data also suggest that the fault comprises several segments with north-south and east-north-easterly trends. East of this fault the Permo-Triassic/Carboniferous boundary is not clearly defined by the gravity data, reflecting the thinning and unconformable overstep of a relatively thin Permo-Triassic sequence on to a gradually shallowing shelf structure.

A second north-north-east-trending low (B), which lies about 15 km east of the Kirkham anomaly, is probably due to a thickening of Permo-Triassic sediments within a basin structure similar to that at Kirkham. The main part of the anomaly lies in the Preston district, just to the south of Figure 25 [around 55 24], and is separated from the northern section by a north-west-trending discontinuity. The east-south-east edge of the anomaly follows the mapped Permo-Triassic/Carboniferous boundary, but to the north-north-

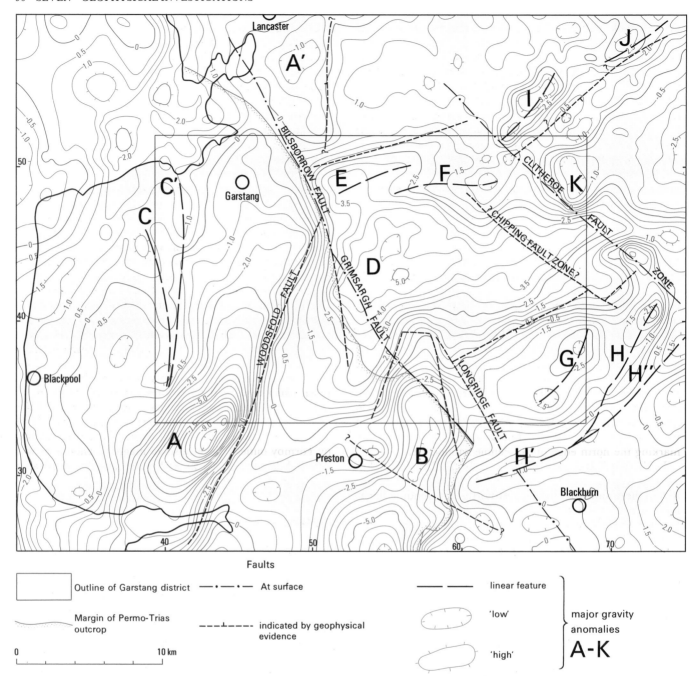

Figure 25 Residual Bouguer gravity anomaly map of the Garstang district and adjoining areas.

extend across the Grimsargh Fault (the mapped Permo-Triassic/Carboniferous boundary) to a possible subsurface fault running subparallel to the Longridge Fault. There is insufficient evidence to determine if these low density rocks are part of the Permo-Triassic basin and, therefore, the origin of the northern part of the anomaly remains problematical.

The trend of minor features in the area of Permo-Triassic rocks tends to be north to north-north-east, mirroring the strike of the two main anomalies. The Weeton Anticline could follow a weak gravity high west of the Kirkham Basin (possibly C or C′).

The centre of the district is dominated by a broad, roughly triangular high (D) which has been interpreted by Lee

(1988a) as a raised basement block lying beneath Dinantian mudstones; it is approximately the same as the South Fells tilt block of Lawrence et al. (1987). Alternatively, if the Dinantian sediments in this area were more calcareous and thus had a higher density, these could produce (or contribute to) the gravity high. The steep gradients which mark most edges of this anomaly indicate that the block is probably fault bounded. Its north-north-west edge lies just north of the Nicky Nook Anticline, which is marked by a weak local high (E), and appears to run from the Bilsborrow Fault eastwards to the Clitheroe Fault System, which is delineated on the gravity map by a series of local lows, steep gradients and interrupted anomalies. Just east of the Nicky Nook Anticline, the Luddocks Fell–Hareden Syncline is marked by a local gravity low (F). The north-east edge of the high is less clearly defined, but could run along the Clitheroe Fault System or, alternatively, along the 'Chipping Fault System' (Lee, 1988a) which runs parallel to, and 4 to 5 km west of the Clitheroe Fault System. However, the 'Chipping Fault System' has not been detected during surface mapping and, if it does exist, it is likely to be a basement structure that has undergone little or no movement since early Carboniferous times.

The south-east edge of block D is also marked by steep gradients, and evidence from seismic traverses shows a north-north-east-trending fault in this area (Thornley Fault, Figure 20), with the thickness of the Carboniferous sequence increasing from about 1 to 4.5 km. The east end of this fault runs into the Clitheroe Fault System, possibly offset southwards along the postulated 'Chipping Fault System', and, to the west, it may continue beneath the Permo-Triassic west of the Longridge Fault. In this area the central high may continue as the broad ridge running south-westwards between the two Permo-Triassic basins (A and B).

East of the Longridge Fault lies a broad east-north-east-trending low, which contains two local north-east-trending minima overlying the Ribchester Syncline (G) and the north limb of the Dinkley Anticline (H). The south-west limit of the Dinkley Anticline is marked by a gravity high (H'), the change from low to high marking the Pendle Grit–Bowland Shale boundary; on gravity evidence, however, this limit would appear to be connected along a ridge to H'', possibly marking the north edge of the Pendle Monocline.

In the north-east part of the map the Catlow and Sykes anticlines are defined by gravity highs (I and J). Just south-east of these anticlines lies an east-north-east-trending lineament which can be traced from the Clitheroe Fault System through the north-east corner of the map into the Settle district. This has been termed the 'Bowland Line' by Arthurton et al. (1988) and interpreted as a possible major block boundary. This lineament may be represented in the Garstang district by the steep gradient (Thornley Fault) on the southern margin of high D, in which case a sinistral displacement of 7 to 8 km across the Clitheroe Fault System is implied. Alternatively, the two lineaments represent the raised south-eastern margins of two blocks that tilted independently on either side of the Clitheroe Fault System. A pronounced local gravity low (K) coincides with the Birkett Fell graben of low density Namurian rocks.

AEROMAGNETIC SURVEY

The aeromagnetic survey of the United Kingdom in 1958 covered the Garstang area with east–west flight lines spaced 2 km apart and north–south tie lines at 10 km intervals. Mean terrain clearance was 305 m (1000 ft). The original analogue maps have subsequently been digitised and this digital data set was used to generate the total field map presented in Figure 26. This shows no major anomalies, the area being dominated by a regional gradient of 2 to 3 nT/km extending north-westwards from a low (−80 nT) centred between Preston and Blackburn to a weak high (+10 nT) near Lancaster. Towards the east side of the map the contour strike becomes more east-north-easterly, the dominant trend of the Carboniferous exposed in the Craven Basin. The western edge of the map is dominated by a low centred around Blackpool.

There is little evidence from the aeromagnetic data that magnetic rocks exist within the Permo-Triassic and Upper Carboniferous sequences. The broadly north-easterly regional magnetic strike extends across the area from the Settle district, in the east, out into the Irish Sea, implying that the magnetic basement rocks have a similar trend or structural control as the surface Carboniferous rocks. Over much of the district the aeromagnetic data suggest that the Carboniferous strata are underlain by nonmagnetic rocks, possibly a Lower Palaeozoic sedimentary sequence. Magnetic basement rocks probably approach the base of the Carboniferous near Lancaster and 20 km east of Blackburn.

In addition to the regional airborne survey, a special high resolution aeromagnetic/radiometric survey has been flown over a small part of the area (A, Figure 26) using high sensitivity equipment to search for hydrocarbon induced micromagnetic and radiometric anomalies. The survey was flown using east–west flight lines spaced 200 m apart; the magnetic and radiometric sensors were at 100 m and 130 m above terrain respectively. This area was chosen as an additional test site following a similar survey flown over the nearby Formby oilfield. Several small magnetic anomalies were detected, but the results generally are inconclusive (Peart et al., 1989).

OTHER SURVEYS

Mineral reconnaissance surveys have been carried out in the Craven Basin by BGS on behalf of the Department of Industry (B and C, Figure 26), some of which involved the use of geophysics. One of the survey areas (B; Cow Ark) lies partly within the Garstang district. The surveys involved the use of induced polarisation and VLF-EM techniques to investigate mineralisation in limestone (p.95); the results are reported fully by Wadge et al. (1983).

Geophysics has also been used to investigate the hydrogeological properties of the Sherwood Sandstone and the drift which covers most of this rock unit. An airborne survey was flown for the Water Resources Board by Barringer Research Limited (1973) to map the resistivity of near-surface deposits. Birmingham University has also conducted a number of geophysical surveys, using resistivity soundings, seismic refraction and detailed gravity to aid geological mapping and to examine drift characteristics and deeper structures for recharge quality and quantity of the aquifer (Barker, 1974; Worthington, 1972; Worthington and Griffiths, 1975).   SFK

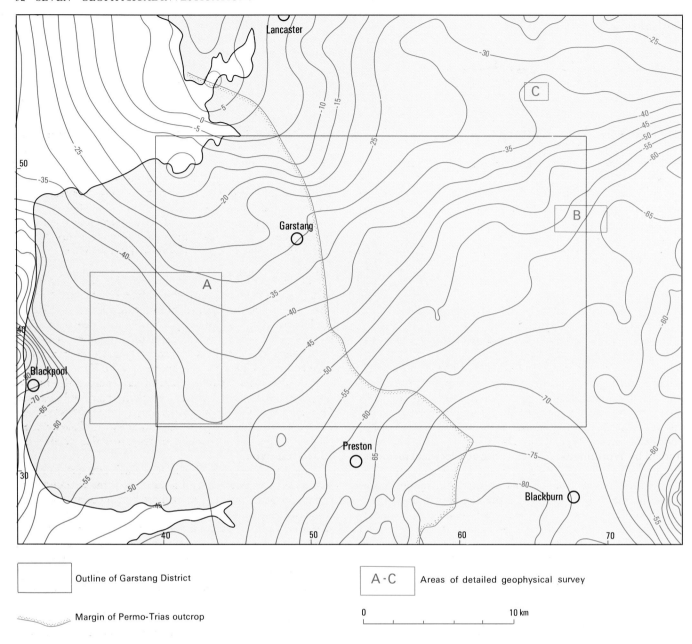

| | Outline of Garstang District | | A-C | Areas of detailed geophysical survey |

Margin of Permo-Trias outcrop

0   10 km

Figure 26   Aeromagnetic map of the Garstang district and adjoining areas. Contours interpolated in areas of anomalies suspected of having non-geological origins.

EIGHT

Economic geology

UNDERGROUND WATER

A major aquifer, the Sherwood Sandstone Group (described in older literature as the 'Bunter Sandstone aquifer'), crops out extensively beneath drift deposits in the district. Over 90 abstraction and monitoring boreholes have been sunk into this aquifer (Figure 14), with additional boreholes drilled on multi-well sites. These wells were mostly sunk for the former Fylde Water Board between 1956 and 1971. Others have been drilled for private and industrial users, including ICI who required water for their operations in the Preesall salt field in the adjacent Blackpool district. As long ago as 1973, Barker and Worthington (1973) described the aquifer as being in an advanced stage of development.

In general, the aquifer is confined by the thick and extensive drift cover. Research on the hydrogeology has been concentrated mainly on trying to understand how surface water migrates through this drift cover into the aquifer to compensate for water abstraction, and to determine how groundwater moves within the aquifer. The three-dimensional distribution of relatively impermeable clayey deposits and permeable sandy deposits in the drift sequence has been studied by Worthington (1972), using geoelectrical methods; the results gave a useful approximation of this distribution.

It is now known from motorway site-investigation boreholes that the drift deposits are more complex than was previously indicated, as can be seen by comparing the drift sections drawn by Worthington (1972, fig. 8) and Sage and Lloyd (1978, fig. 6) with Figure 18 in this memoir. Nevertheless, as Sage and Lloyd have concluded from a hydrochemical study, bodies of sand or sand and gravel, though irregular in shape and laterally impersistent, probably do provide an important route for surface water to pass through the drift to recharge the aquifer. Another probable recharge route revealed during the present survey is via exposures of Sherwood Sandstone in the beds of the River Wyre [4967 4856] near Scorton and the River Brock [4929 4024] at Myerscough. Whether recharge or discharge of the aquifer takes place at these localities will depend critically on the level of the water table, which in turn depends on the rainfall and the amount of water abstracted from wells. The presence of Sherwood Sandstone near or at the ground surface at three localities [4634 5065; 4571 4986; 462 507] near Great Crimbles probably explains the discharges from the aquifer into the River Cocker that are thought to have occurred in the extreme drought of 1976 (Sage and Lloyd, 1978, p.217).

The level of the water table represented by 'piezometric contours' shows a general fall from east to west (Sage and Lloyd, 1978, fig. 4), as might be expected from the general fall in ground level in the same direction . There are, however, deviations in detail from this general trend close to the line of the Bilsborrow Fault. These indicate that the water table is also affected by recharge from sandy drift deposits and from Carboniferous sandstones along the fault line. Oakes and Skinner (1975) reported that aquifer tests carried out in wells in the Sherwood (Bunter) Sandstone close to this fault show hydraulic continuity along the fault line in the Catteral area. Sage and Lloyd (1978, fig. 5) also show that there are complex variations in the distribution of relatively hard and soft waters within the aquifer, as illustrated by the distribution of bicarbonate ion concentrations. They suggest that zones of softer water are located where indirect recharge by river water occurs via sandy drift deposits.

The probability also exists that the Sherwood Sandstone aquifer is recharged from groundwater held in Millstone Grit Group sandstones brought into contact with it along the Bilsborrow Fault (Figure 21). Clearly, the extent to which such recharge occurs will depend on a number of factors, particularly the extent of the contact at the fault plane or fault zone, and the aquifer properties of the Namurian sandstones. The latter factor appears to have been little studied, only the Pendle Grit having been penetrated by boreholes. Of the five boreholes concerned, only two (SD 54 NW/1 and 2, drilled to supply water to a factory at Oakenclough) tapped a significant thickness of the formation (208 m and 106 m respectively).

Although the Pendle Grit is very thick and has an extensive outcrop, it does contain a number of interbedded shaly sequences that would reduce its bulk permeability, especially where these are laterally persistent. Sandstones higher in the Millstone Grit succession in the Scorton and Barnacre areas, notably the Park Wood Sandstone, probably contain a much higher proportion of argillaceous beds. On the other hand, the Stonehead Sandstone has a massive well-jointed appearance (p.53) that suggests it might be a good aquifer if the joints are laterally persistent.

The geological map indicates that the most likely locations where the Millstone Grit sandstones will be in contact with the Sherwood Sandstone extend along the Bilsborrow Fault line from near Forton [479 519] to Parkhead [507 455], and from Duckett's Farm [513 417] to Anderton Fold [515 398]. The latter tract consists entirely of the conjectural downfaulted subdrift outcrop of the Pendle Grit (Figure 20). This outcrop is isolated from the main Pendle Grit outcrop by the Grimsargh Fault. The fault may restrict the movement of water from the large groundwater catchment area of the Bowland Fells. There is, however, some evidence for hydraulic continuity between the Sherwood Sandstone aquifer and this downfaulted block of Pendle Grit between the Grimsargh and Bilsborrow faults (see below).

In the south-east part of the district, the Grimsargh Fault throws the combined Pendle Grit and Warley Wise Grit outcrops against the Sherwood Sandstone between Grimsargh [587 343] and near Chapel House Farm [596 332], to the east of Grimsargh Hall. Again, however, this is a faulted tract which mapping suggests may not be in hydraulic continuity with the extensive outcrop of the Pendle and Warley

Wise grits forming the groundwater catchment area of Longridge Fell. It should be emphasised that, both here and in the areas mentioned above, the fault lines and the adjacent sandstone outcrops lie beneath a thick drift cover; their positions are therefore conjectural.

While recharge of the aquifer by fresh water is clearly desirable and beneficial, the reverse is true for contamination by saline waters. There are two potential sources of contamination, namely Morecambe Bay sea water in the northwest of the district and groundwaters in the salt-bearing formations in the Mercia Mudstone Group. For this reason, abstraction boreholes tend to be sited away from both the coast and the Mercia Mudstone outcrop. Pumping in the few boreholes near the coast has to be carefully monitored and controlled, otherwise the natural outflow of groundwater towards Morecambe Bay could be reversed (Oakes and Skinner, 1975, pp.18–19). Prolonged overpumping would result in saline water being drawn into the aquifer. Sage and Lloyd (1978, p.218) imply that this danger is not as great as might be expected because analyses of saline water from wells that have been contaminated do not indicate a modern origin. They further suggest that 'thick drift deposits form a reasonably effective confining layer'. It should be noted, however, that recent BGS mapping (Crofts, 1987) records exposures of well-jointed Sherwood Sandstone in the intertidal zone on the northern edge of Cockerham Sands [427 533] only about a kilometre north of the district boundary. These exposures could readily allow ingress of sea water into the aquifer.

The way that water moves through the Sherwood Sandstone aquifer has been a matter of some debate. Oakes and Skinner (1975) showed that marked variations in transmissivity occur over much of the area. The possibility that flow through fissures in the sandstone might be the dominant mechanism was investigated by Brereton and Skinner (1974). They showed that, for the one borehole they investigated in detail (SD 44 SE/28), water inflow was largely through a variety of fissures that seemed to be of limited extent and multidirectional orientation. They inferred from generally available data that intergranular permeability is relatively uniform over the Fylde area and that large variations in fissure development accounted for corresponding variations in borehole yields. Worthington (1977) came to the opposite conclusion, namely that there are marked spatial variations in the intergranular component of groundwater flow and that it was this factor that 'exerted a basic control on abstraction well productivity.'

There are now indications from seismic data that the Sherwood Sandstone was subjected to great tensional stresses during and shortly after deposition, resulting in growth faulting and large thickness variations (see p.86). The effects of this early episode in the aquifer's structural history are not known in detail but seem certain to have produced considerable inhomogeneities in sandstone permeability, both of the fissure-associated and intergranular types.

SAND AND GRAVEL

1:10 000 scale geological mapping has indicated that numerous significant outcrops of sand, and sand and gravel,

exist in the district in the Quaternary drift deposits. These include Glacial Sand and Gravel, Glaciolacustrine Sand, Older Storm Beach Deposits, River Terrace Deposits (including terrace deposits of the River Wyre), Alluvial Fan Deposits and Blown Sand. In many instances, the full extent of these deposits is largely concealed, either beneath the most recent spreads of river and estuarine alluvium or within the till sheet in the case of the Glacial Sand and Gravel deposits. With further detailed exploration, however, it seems probable that new resources will be discovered.

To date, the main exploitation has been in connection with the construction of the M6 motorway in the early 1960s. Three deposits were worked: Glacial Sand and Gravel at Bradley's Sand Pit [5110 3400] near Broughton (Plate 17), Alluvial Fan Deposits of the River Calder in a borrow pit [514 432] near Claughton, and alluvial floodplain gravels of the River Wyre upstream from the railway bridge [5952 4937] near Scorton. At the time of writing, in 1989, the first and last of these workings were still in production. Exploitation of the Calder alluvial fan deposit ceased after completion of M6 motorway construction, probably because of the high silt and clay content of the poorly sorted gravels.

The Scorton deposit is currently being worked by Tarmac Roadstone Limited, and this company have also opened up a newly discovered deposit of probable glacial origin at Headnook Farm [505 388] on the Myerscough Estate, south-west of Bilsborrow.

There are no other workings in the district except for the minor diggings which are commonly present in the more obvious outcrops.

LIMESTONE

Though no limestones are quarried today in the district, they were extensively exploited in the past, mainly for the production of agricultural lime, road aggregate and locally used building stone. Numerous disused quarries in the main limestone formations, and the ruins of small limekilns, remain as evidence of this former exploitation (Plate 3). Although the limestones are similar to those currently being extensively quarried around Clitheroe, their exploitation is discouraged by the lack of good road and rail links and by the designation of the area as one of 'outstanding natural beauty'.

An assessment of the limestone resources of the Craven Basin to the east of the Garstang district was carried out by BGS in 1979–81 (Harrison, 1982*). The report has some relevance to the present district because the facies of the Chatburn Limestone and the knoll-reef limestones within the Clitheroe Limestone Formation, in particular, change little across the basin. The report states that the knoll-reef limestones have the highest purity (>97 per cent $CaCO_3$) and include some of chemical grade. The purity of the other limestone units varies mainly according to the amount of silica present, either as discrete chert bodies or in disseminated form, and the proportion of argillaceous sediment interbedded with the limestones.

---

* The inclusion of sheets 59 and 67 in the title of the report referred to was erroneous.

SANDSTONE

Numerous disused quarries on Longridge Fell in the massive, thickly bedded and generally well-cemented sandstones of the Pendle Grit Formation attest to its former value in the region as a building stone. The group of quarries immediately east of the town of Longridge are said to have produced no less the 30 000 tons of stone annually in its heyday (De Rance, unpublished m/s, 1888). Today, only Leeming Quarry [683 406] (Brown Bros.), at the eastern margin of the district, is still in production (Plate 14). This quarry is able to supply large blocks of dressed stone similar in quality and appearance to the Longridge Stone that has been used in a number of major buildings including Preston Town Hall. Smaller disused quarries, presumed to have worked stone for mainly local building purposes, are present on Beacon Fell and in the Pendle Grit outcrop west of the high Bowland Fells. Two of the sandstones in the Roeburndale Formation, namely the Park Wood Sandstones and the Stonehead Sandstone, have also been worked in small quarries around Barnacre.

Pendleside Sandstones have been quarried locally around Chipping for building stone. Variable bed thicknesses and the presence of shaly interbeds have probably restricted the use of these sandstones.

Flaggy sandstones in the Sabden Shales at Buckley Hall, near Ribchester, were formerly worked for use as roofing stone. Here [6400 3660], as well as overgrown disused quarries, there are signs of shafts and a collapsed adit, indicating that the flagstones were also mined.

BRICK CLAY

It is likely that clays from the glacial tills have been used in the past for brick, tile and pipe manufacture, but little evidence remains of this industry. There are numerous overgrown pits in the district, but the great majority of these were probably dug to improve the agricultural quality of the land by 'marling' (Hall and Folland, 1970, p.90). Pits that almost certainly were dug for brick, tile and pipe making include those around Crimpton and Spire Farm [6800 4675; 6791 4706; 6807 4662]; other examples are listed by Fletcher (1990).

PEAT

In the Fylde the deposits of peat, in places over 3 m thick, that formerly extended over large areas known as 'mosses' have been greatly reduced by exploitation for fuel; some peat cutting is still carried out on a small scale, for example south of Tarn Farm [448 489] Hill peat on the Bowland Fells and Longridge Fell has also been dug but is no longer an economical source of fuel.

METALLIFEROUS MINERAL DEPOSITS

There is evidence for some mineralisation in the north-east part of the district but it is of only minor economic signifi-

cance. This part of the Craven Basin lies some 22 km from the Craven Fault Belt (Figure 20), the line that marks the south-west margin of the nearest area of major mineralisation, the Northern Pennine Orefield (Dunham and Wilson, 1985). These authors concluded that the lead-zinc-fluorite-baryte deposits were hydrothermally emplaced in the Carboniferous limestone and sandstone host rocks, crystallising out of hot brines. The fluids were largely of connate origin and had been expelled from penecontemporaneous basinal mudstones during deep burial and under tectonic stress, mainly during late Carboniferous and Permian times. Plant and Jones (1989) have recently suggested a similar origin for the Pennine Orefields mineralisation farther south, which they describe as a modified, tectonically driven Mississippi Valley type (MVT) ore deposit.

The primary sulphides, mainly pyrite (FeS_2), galena (PbS) and sphalerite (ZnS), deposited by the process outlined above, were subsequently partly altered by oxygen-bearing groundwaters, probably assisted by microbial action (Ehrlich, 1981). There was then further alteration and remobilisation, and a suite of secondary minerals was deposited (see below).

The relatively minor deposits in the present district probably have a similar genesis to those in the main orefield. Dinantian limestones form the host rocks, particularly the more massive and thicker bedded members of Chadian or Arundian age. The distribution of these limestones, and four locations of associated mineralisation, are shown in Figure 27. Lead and zinc minerals have been worked at three of these locations, namely Sykes, Moor End and Dinkling Green. At the fourth, Spire Farm near Cow Ark, there is no record or evidence of old workings, but there was some intensive exploration by BP Minerals International Ltd in 1983, largely as a result of a basinwide BGS mineral reconnaissance survey in 1973, funded by the Department of Industry (Wadge et al., 1983). BP Minerals has also carried out exploration at Sykes.

Locally, notably in the Sykes Anticline, which just impinges on the northern margin of the district, the limestones are silicified and dolomitised (see below). The silicification is stratigraphically transgressive but it is not clear from field evidence whether it was synchronous with lead-zinc mineralisation (Hughes, 1986). Dolomitisation is stratabound and is mainly confined to the Pendleside Limestone, in some places affecting only certain beds or groups of beds within this formation. General observations on dolomites at this level in the Craven Basin indicate that the dolomitisation process postdated major compaction and may have taken place during Silesian times, in association with a remagnetisation event (Addison et al., 1985). These authors also favour 'mudrock-dewatering' as a mechanism for generating the dolomitising fluids.

Details

Sykes

Mining at Sykes may have been taking place as long ago as 1768 when a small amount of lead ore, probably from this locality, was smelted at Grassington (Gill, 1987, p.43). The mine was probably not as productive or extensive as that at Brennand in the adjacent Lancaster district, however. De Rance (1873) has described the

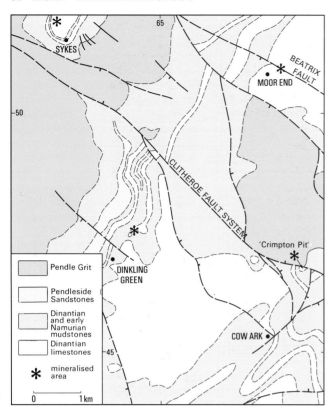

Figure 27 Location and general geological setting of mineralised areas in the north-east of the district.

geology and mine workings in some detail, but mining appears to have almost ceased when he made his survey. A plan and section of the workings made by Earby Mines Research Group and published by Gill (1987, fig.16) accords fairly closely with De Rance's description but shows no geological details. The main orebody seems to have been an east-north-east-trending vein, with some irregular off-shoots, following a minor fault plane on either side of the valley of the Losterdale Brook [6280 5183] and lying subparallel to the major axial trace of the Sykes Anticline. The host rock is the Hetton Beck Limestone (see p.25), which has also been quarried at this locality. De Rance (1873) notes that the slickensided footwall of the vein hades south at 52°, and Moseley indicates that the fault had a small reverse throw to the north-north-west. Galena, barytes and green copper staining are common in the waste tips. Quartz, fluorspar and calamine have also been recorded by Raistrick (1973, p.163).

The Hetton Beck Limestone is extensively silicified in its upper part, immediately overlying the Sykes mine workings. The silicification is transgressive, also affecting strata overlying the Hetton Beck Limestone, including the Hodder Mudstones, the Hodderense Limestone and the Pendleside Limestone, on both flanks of the Sykes Anticline north-north-west and south-south-east of the mine workings. Some dolomitisation is also present, mainly affecting the Pendleside Limestone.

Moor End

East-north-east of Moor End a vein marked by old surface diggings appears to lie along the line of the Beatrix Fault [6751 5091 to 6765 5082]. According to Fletcher (1991), calcite is abundant and, in places, lines vugs in the Thornton Limestone, but no metalliferous minerals are evident. Another group of pits lies to the north [6757 5111] and farther along strike to the north-east, but there is no sign of mineralisation. Historical records summarised by Gill (1987)

suggest that the mines had a long history and at one time produced a significant quantity of silver.

Dinkling Green

A few minor workings in the well-exposed slopes of the knoll-reefs near Dinkling Green were said to be exploiting small "sops" or pockets of the zinc carbonate ore smithsonite, commonly known as calamine (De Rance, 1873, p.73). The largest such working is shown on earlier editions of Ordnance Survey maps at a scale of six inches to one mile as a 'calamine pit' [6442 4753]. Gill (1987) describes the workings in the vicinity as having a vertical range of about 23 m and as forming an interconnecting series of tight passages, inclines and shafts. He also notes that these workings are similar to those at Ashnott, also in knoll-reef limestones, situated just beyond the boundary with the Clitheroe district, 5 km to the east.

The knoll-reefs originated as carbonate mudmounds in which a phreatic cavity system formed at an early stage in diagenesis (Miller, 1986). The irregular form of the minor orebodies, as indicated by the complexity of the workings, probably reflects an intricate system of cavities. Miller (1986) has shown that the cavities had a long and complex history of infilling with carbonate cements, and it seems likely that some cavities may have been reopened by dissolution and penetrated by metalliferous hot brines, with subsequent sulphide emplacement. If, as seems likely, the brines emanated from the overlying and enclosing mudstone sequence, those knoll-reefs which had a markedly discordant unconformable contact with the mudstones would have been favoured sites for sulphide mineralisation. The transformation to secondary minerals such as smithsonite, which seems to have occurred at higher levels in the knoll-reefs (De Rance, 1873, p.74), probably resulted from groundwater circulation beneath the Permian or Triassic land surface.

Cow Ark

Mineralisation associated with a knoll-reef in the Clitheroe Limestone Formation, located [6799 4677] 1.5 km north of Cow Ark, was noted briefly by Poole (in Earp et al., 1961, p.43). The immediate area was explored by Wadge et al. (1983) in 1976–77 as part of the Mineral Reconnaissance Programme funded by the Department of Industry. Various techniques were used, including geochemical soil sampling, induced polarisation (IP) and very low frequency-electromagnetic (VLF-EM) surveys. These were followed by a programme of vertical and inclined drilling in which ten cored boreholes were sunk to depths of up to 220 m. The results of this work were sufficiently encouraging to stimulate BP Minerals Development Ltd to carry out further exploration in the area. No exploitable reserves were found, however. There is no record of mining in the Cow Ark area prior to this recent exploration.

The geological map published by Wadge et al. (1983, fig.58) was revised during the recent survey in the light of the new exploration boreholes drilled by BP Minerals and further biostratigraphical work on the BGS borehole cores (p.13). Crimpton Pit [6799 4677], where galena, sphalerite and, more commonly, smithsonite occur, is excavated mainly in glacial till overlying altered knoll-reef limestone at the core of a dome-like structure truncated by the Clitheroe Fault. This fault throws down the basal interbedded sandstones and mudstones of the Pendle Grit Formation on its north (hangingwall) side. The few boreholes that intersect the fault plane suggest that no particular concentration of minerals is associated with this structure. Rather, such mineralisation as occurs is found where limestones with some primary or secondary porosity are overlain by relatively impermeable mudstones. The main units are the Limekiln Wood and Pendleside limestones, and the flank and toe facies of the knoll-reef limestones. The latter also locally retained enough porosity to allow the entrapment of small quantities of oil (Plate 16).

Wadge et al. (1983) found that pyrite is by far the commonest sulphide present in borehole cores; lead and zinc sulphides occur only in minor quantities. Like the mineralisation at Dinkling Green (see above), secondary minerals, notably smithsonite or calamine, are most common in the surface outcrops.

Other occurrences

Minor calcite and baryte veins occur in most limestones exposed in the district. They are associated particularly with faults and the axial culminations of minor folds.

Anomalously high stream and soil values for Pb, Zn, Cu and Ba have been detected in a 6 km² tract around the valley of the River Loud near Little Elmridge [599 415] (Wadge et al., 1983, fig.75). The anomalies lie mainly on thin glacial lodgement till overlying a sequence comprising the Pendleside Sandstones and overlying mudstones and subordinate argillaceous limestones of the Bowland Shale Group. Most of the anomalous values lie within 200 m of probable fault lines, notably the Oakenclough Fault [580 430 to 603 415], lending support to the conclusion of Wadge et al. (1983) that minor sulphide enrichment had taken place along these faults.

HYDROCARBONS

The presence of oil and gas in the Carboniferous rocks of north and central England has long been known or suspected (Plate 16). The nearest places where production has been achieved are the Formby oilfield (Falcon and Kent, 1960) and the offshore Morecambe gas field (Levison, 1988). These localities lie some 30 km to the south-south-west and 45 km to the west of the district respectively. The source rocks in the latter field are thought to be Coal Measures strata, which are unlikely to occur around Garstang, though their presence beneath the Permo-Triassic outcrop cannot be entirely ruled out. The source rocks at Formby, which are late Dinantian to early Namurian marine shales, are present in both the exposed and concealed Carboniferous outcrops in the district. It is likely that concealed source rocks of early Dinantian age are present also, but there are no published data and the sequence is largely unproven.

The generation of oil and gas in source rocks depends on high temperature which, in turn, depends mainly on depth of burial (Tissot and Welte, 1984, fig.II.7.1.). The accumulation of an economic resource basically depends on whether the oil and gas are able to migrate to a porous reservoir rock situated in a geological structural trap that will prevent further migration and escape.

Fraser et al. (1990) have recently provided a summary of critical factors for the assessment of a possible oil or gas resource in the Carboniferous and associated rocks of northern England. They are as follows:

(1) critical burial depth for preservation of reservoir quality (c.2550 m).
(2) critical burial depth for the oil window (1900–3600 m).
(3) distribution of good quality basinal pro-delta source rocks.
(4) degree of post Variscan (Hercynian) trap modification.

These authors cite the Bowland and Sabden shales as being amongst the 'best quality oil-prone source rocks'. Since

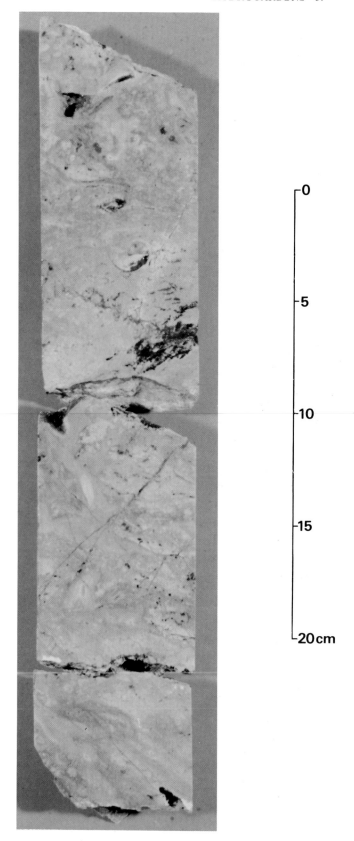

Plate 16   Slabbed core of Waulsortian Limestone from the Bellman Limestone Member (early Chadian), showing oily hydrocarbon impregnation into geopetal cavities and in later ferroan dolosiltstone patches. Borehole SD 64 NW/2, 13.22 to 13.37 m depth.

these formations occur in the district, probably beneath the Permo-Triassic cover in the west as well as at outcrop in the east, critical factor (3) is satisfied. Concerning critical factor (2), Metcalfe et al. (in press) have estimated the thickness of the Silesian sequence to have been in the order of 3500 m, comprising 1500 m of Westphalian and 2000 m of Namurian strata. This implies that the Bowland Shales (latest Dinantian – earliest Silesian) were buried at least to this depth, and probably more if compaction effects are allowed for, since the thicknesses estimated are for rocks that have already been compacted. Thus burial depths were probably near or greater than the 3600-metre 'oil window' lower depth limit quoted above, but well within the gas generation depth range. Conodont alteration indices (CAI) from Dinantian limestones in the Craven Basin, together with a limited amount of vitrinite reflectance data (Lawrence et al., 1987), led Metcalfe et al. (in press) to conclude that post-Carboniferous burial by Mesozoic rocks had little temperature-enhancing effect. In the west, however, growth faulting during Permo-Triassic sedimentation greatly increased the thickness of the accumulating sediments (Figure 24) and reburied the source rocks, probably to depths where oil and gas generation was resumed (Fraser et al., 1990, fig.24). The youngest thick Namurian marine mudstone sequence, the Caton Shales, lies about 1000 m above the basal Namurian Bowland Shales level and would therefore have a burial depth of c.2600 m, i.e. near the oil window maximum (2700 ± 200 m). Other marine mudstone sequences of possible source rock potential, i.e. those associated with the C. cowlingense and E. ferrimontanum marine bands lie at intermediate levels (Figure 10), but are probably too thin to be significant.

Sandstones which lie above the critical burial depth of c.2500 m for the preservation of reservoir quality (critical factor (1) above) would include the little known Heversham House, Wellington Crag and Ellel Crag sandstones. These are present only in the extreme north of the district where they crop out and are therefore not involved in a structural trap. Critical factor (1) is based on core data from sandstones of delta-top fluvial channel facies in the East Midlands, which are more likely to resemble the cross-bedded Wellington Crag and Ellel Crag Sandstones than the turbiditic Heversham House Sandstone. These and older Namurian sandstones are likely to occur beneath the Permo-Triassic outcrop in the west of the district, but their depth, facies and structural geometry is unproved.

Two Mesozoic sandstones are known to have reservoir potential in the region, namely the early Permian Collyhurst Sandstone and the uppermost part of the Triassic Sherwood Sandstone (Keuper Waterstones), both in the Formby oilfield (Falcon and Kent, 1960). The latter formation is also the reservoir in the Morecambe gas field (Levison, 1988). In the western part of the district the local presence of the former reservoir is likely but not proven, whereas the latter is certainly present though not necessarily oil or gas bearing.

The Hercynian (Variscan) compressional or transpressional fold movements (p.86) provide the principal trap-forming mechanism for the Carboniferous rocks in this district, as elsewhere in northern and central England. The north – south-oriented penecontemporaneous rifting, which has markedly affected the Permo-Triassic rocks, may well have had a deleterious effect on some of the pre-existing reservoir traps (critical factor (4) above). On the positive side, however, these rift movements are likely themselves to have produced traps, such as rollover anticlines, in any Mesozoic reservoir rocks that were present.

The considerations outlined above indicate that detailed hydrocarbon exploration in the region that includes the Garstang district is probably justified. It is not surprising therefore that exploration licences have been taken by several companies, which currently include British Gas, Enterprise and Pendle Petroleum. Many kilometres of seismic traverse have been made as well as sampling and analysis of possible source and reservoir rocks from the exposed outcrops. In December 1988, the first 'wildcat' well in the district was drilled by British Gas, and others are planned. The results of most of this exploration effort have yet to be released, but they are certain to provide a great increase in our knowledge and understanding of the geology of the district.

POTENTIAL GEOLOGICAL RISK FACTORS

This necessarily brief account is provided to supplement the obvious sources of potential hazard to engineered structures, such as large landslips and peat deposits. Geological maps, which show the extent of these deposits, give the essential first indication of ground conditions and help to enable site investigations to be planned in the most efficient and effective way. It is emphasised that BGS maps, even at 1:10 000 scale, do not provide information detailed enough for site investigation purposes, and that site specific investigations should always be carried out prior to any development. Unlike many urban areas, for which BGS has a large geotechnical database consisting largely of site investigation borehole records, there is little geotechnical information available for the present district. However, certain hazards associated with particular mapped deposits are mentioned briefly below.

Till

Although glacial till is generally a stiff sandy clay deposit, it may contain concealed lenses of sand with varying proportions of gravel, which can be highly charged with water. In some areas, such as near New Draught Bridge [4779 4010] and notably at several localities in the valley of the River Hodder [645 505; 654 485; 645 433], partially or wholly concealed laminated clays occur in the till and beneath alluvial deposits, and these can weaken the associated deposits and add to the instability caused by river erosion. Landslipping has been intermittently active at all three of the Hodder Valley localities in recent years, with potential or actual damage to roads in the vicinity.

Many of the till-covered steep valley sides in the eastern part of the district are potentially unstable, although no actual clearly defined slip has been mapped. Additionally, solifluction and gravitational creep will almost certainly have intermittently mobilised at least the uppermost metre or so of the till sheet, perhaps imparting small-scale shear planes within the fabric and thereby reducing its strength. Strictly, this mobilised till layer should be classified as head, but since it is hardly distinguishable from the parent material, it has not been separately delineated on the maps.

Peat

In addition to the clearly marked spreads of peat on the low-lying areas of the Fylde and on the high moorlands, there are also concealed peaty deposits in certain locations. These include peat, and silt or sand rich in organic matter. They may occur as channel fills within alluvial and river terrace deposits, as beds within estuarine and lacustrine clays, silts and sands, and in kettle holes on the till surface. Such peat layers would be expected to be of low strength, high compressibility and acidic water content. Because some bodies of peat, such as those formed in kettle holes, are laterally restricted, they can result in severe differential settlement of building foundations if not removed before construction.

Organic deposits can also generate methane, and their burial may introduce a risk of explosive gas concentration.

Methane in groundwater

The underground explosion at Abbeystead, just north of the district, which resulted in the tragic death of sixteen local people on 23 May 1984, focused attention on the dangers of methane dissolved in groundwater (Health and Safety Executive, 1985). Carboniferous marine mudstones, which are source rocks for hydrocarbons including methane, underlie much of the district and form extensive outcrops in the east. The fact that methane is soluble in groundwater, especially under pressure, creates a potential explosion hazard in any structure built or founded in the Carboniferous rocks, or in younger rocks likely to be in hydraulic continuity with the Carboniferous rocks. Therefore, designs for such structures should always allow for free air ventilation.   NA

REFERENCES

Most of the references listed below are held in the Library of the British Geological Survey at Keyworth, Nottingham. Copies of the references can be purchased from the Library subject to the current copyright legislation.

ADDISON, F T, TURNER, P, and TARLING, D H. 1985. Magnetic studies of the Pendleside Limestone: evidence for remagnetization and late-diagenetic dolomitization during a post-Asbian normal event. *Journal of the Geological Society of London*, Vol. 142, 983–994.

AITKENHEAD, N. 1977. The Institute of Geological Sciences borehole at Duffield, Derbyshire. *Bulletin of the Geological Survey of Great Britain*, No. 59, 1–38.

— 1990a. Geology of the Chipping area. *British Geological Survey Technical Report*, WA/90/35.

— 1990b. The geology of the Beacon Fell area. *British Geological Survey Technical Report*, WA/90/52.

ALLEN, J R L, and WRIGHT, V P. 1989. Paleosols in siliclastic sequences. *Postgraduate Research Institute for Sedimentology, Reading University Short Course Notes,* No. 001.

ARTHURTON, R S. 1984. The Ribblesdale fold belt, NW England — a Dinantian–early Namurian dextral shear zone. 131–138 in Variscan tectonics of the North Atlantic Region. HUTTON, D W H, and SANDERSON, D J (editors). *Special Publication of the Geological Society of London*, No 14.

— JOHNSON, E W, and MUNDY, D J C. 1988. Geology of the country around Settle. *Memoir of the British Geological Survey,* Sheet 60 (England and Wales).

AUDLEY-CHARLES, M G. 1970. Triassic palaeogeography of the British Isles. *Quarterly Journal of the Geological Society of London*, Vol. 126, 49–89.

BARKER, R D. 1974. A gravity survey of northwest Lancashire. *Geological Journal*, Vol. 9, 29–38.

— and WORTHINGTON, P F. 1973. The hydrological and electrical anisotropy of the Bunter Sandstone of northwest Lancashire. *Quarterly Journal of Engineering Geology*, Vol 6, 169–175.

BARNES, B. 1975. Palaeoecological studies of the late Quaternary period in north west Lancashire (Fylde). Unpublished PhD thesis, University of Lancaster.

BARRINGER RESEARCH LIMITED. 1973. Report on a ninth-frequency E-phaser' survey of the Garstang area, Lancashire, UK., for the Water Resources Board. (Toronto: Barringer Research Limited.)

BISAT, W S. 1924. The Carboniferous goniatites of the north of England and their zones. *Proceedings of the Yorkshire Geological Society*, Vol. 20, 40–124.

— 1928. The Carboniferous goniatite-zones of England and their continental equivalents. *Compte Rendu Congrès International de Stratigraphie et de Géologie du Carbonifère* (Heerlen, 1927), 117–133.

— 1934. The goniatites of the *Beyrichoceras* Zone in the North of England. *Proceedings of the Yorkshire Geological Society*, Vol. 22, 280–309.

— 1952. The goniatite succession at Cowdale Clough, Barnoldswick, Yorkshire. *Transactions of the Leeds Geological Association*, Vol. 6, 155–181.

BLACK, W W. 1952. The origin of supposed tufa bands in Carboniferous reef limestones. *Geological Magazine*, Vol. 75, 736–738.

— 1954. Diagnostic characters of the Lower Carboniferous knoll-reefs in the north of England. *Transactions of the Leeds Geological Association*, Vol. 6, 262–297.

BOUMA, A H. 1962. *Sedimentology of some flysch deposits; a graphic approach to facies interpretation.* (Amsterdam and New York: Elsevier.)

BOULTON, G S. 1972. Modern Arctic glaciers as depositional models for former ice sheets. *Journal of the Geological Society of London*, Vol. 128, 361–393.

— and PAUL, M A. 1976. The influence of genetic processes on some geotechnical properties of glacial tills. *Quarterly Journal of Engineering Geology.* Vol. 9, 159–194.

BRANDON, A. In preparation. The geology of the Brennand Fell area. *British Geological Survey Technical Report.*

BRAUCKMANN, C. 1973. Kulm Trilobiten aus dem Kulm von Aprath (Bergisches Land). Inaugural Dissertation Freie University, Berlin, Vol. 1, 1–209.

BRAY, A. 1927. The Carboniferous sequence between Lothersdale and Cowling (Colne). *Journal of the Manchester Geological Association*, Vol 1, 44–57.

BRERETON, N R, and SKINNER, A C. 1974. Groundwater flow characteristics in the Triassic sandstone in the Fylde area of Lancashire. *Water Services*, August 1974 issue.

BRIDGE, D. McC. 1988a. Geology of the area around Hurst Green and Wilpshire. *British Geological Survey Technical Report*, WA/88/43.

— 1988b. Geology of the area around Goosnargh and northeast Preston. *British Geological Survey Technical Report*, WA/88/42.

— 1988c. Geology of the area around Barton and Fulwood. *British Geological Survey Technical Report*, WA/88/40.

— 1989. Geology of the area around Longridge and Ribchester. *British Geological Survey Technical Report*, WA/89/66.

BURTON, P W, MUSSON, R M W, and NEILSON, G. 1984a. Marcroseismic reports on historial British earthquakes IV: Lancashire and Yorkshire. *Reports of the British Geological Survey Global Seismology Unit*, Nos. 219a and 219b (2 volumes).

— — — 1984b. Studies of historial British earthquakes. *Report Global Seismology Unit, British Geological Survey*, No. 237.

CHADWICK, R A. 1985. Seismic reflection investigations into the stratigraphy and structural evolution of the Worcester Basin. *Journal of the Geological Society of London*, Vol. 142, 187–202.

— 1986. Extension tectonics in the Wessex Basin, southern England. *Journal of the Geological Society of London*, Vol. 143, 465–488.

CHARSLEY, T J. 1984. Early Carboniferous rocks of the Swinden No. 1 Borehole, west of Skipton, Yorkshire. *Report of the British Geological Survey*, No. 84/1, 5–12.

CHISHOLM, J I, CHARSLEY, T J, and AITKENHEAD, N. 1988. Geology of the country around Ashbourne and Cheadle. *Memoir of the British Geological Survey*, Sheet 124 (England and Wales).

CONIL, R, LONGERSTAEY, P J, and RAMSBOTTOM, W H C. 1980. Materiaux pour l'etude micropaleontologique du Dinantien de Grande-Bretagne. *Memoirs de l'Institut Géologique de l'Université de Louvain*, Vol. 30, 1–187.

CROFTS, R G. 1987. Geological notes and local details for 1:10 000 Sheet SD 45 SW, Cockersand Abbey. *British Geological Survey Technical Report*, WA/87/38.

DAVIES, J R, RILEY, N J, and WILSON, D. 1989. The distribution of Chadian and earliest Arundian strata in North Wales: implications for Dinantian (Carboniferous) lithostratigraphy and palaeogeography. *Geological Journal*, Vol. 24, 31–47.

DAVISON, C. 1924. *A history of British earthquakes*. (Cambridge: Cambridge University Press.)

DEAN, V. 1950. The age and origin of the Sale Wheel gorge and of the deviation of Starling Brook in the same locality. *Journal of the Manchester Geological Association*, Vol. 2, 1–6.

DE RANCE, C E. 1873. On the occurrence of lead, zinc and iron ores in some rocks of Carboniferous age in the north-west of England. *Geological Magazine*, Vol. 10, 64–74.

— 1877. The superficial geology of the country adjoining the coasts of southwest Lancashire. *Memoir of the Geological Survey of Great Britain*.

DUNHAM, K C, and WILSON, A A. 1985. Geology of the Northern Pennine Orefield: Volume 2, Stainmore to Craven. *Economic Memoir of the British Geological Survey*.

DUNHAM, R J. 1962. Classification of carbonate rocks according to depositional texture. *Memoir American Association of Petroleum Geologists*, No. 1, 108–121.

EARP, J R. 1955. The geology of The Bowland Forest Tunnel, Lancashire. *Bulletin of the Geological Survey of Great Britain*, No. 7, 1–12.

— MAGRAW, D, POOLE, E G, LAND, D H, and WHITEMAN, A J. 1961. Geology of the country around Clitheroe and Nelson. *Memoir of the Geological Survey of Great Britain*, Sheet 68 (England and Wales).

— and TAYLOR, B J. 1986. Geology of the country around Chester and Winsford. *Memoir of the British Geological Survey*, Sheet 109 (England and Wales).

EHRLICH, H L. 1981. *Geomicrobiology*. (New York: Marcell Drekker, Inc.)

EVANS, W B, and WILSON, A A. 1975. Outline of geology on Sheet 66 (Blackpool) of 1:50 000 Series: Geological Survey of Great Britain.

EYLES, N, and McCABE, A M. 1989. The Late Devensian (<22 000 BP) Irish Sea Basin: the sedimentary record of a collapsed ice sheet margin. *Quaternary Science Reviews*, Vol. 8, 307–351.

FALCON, N L, and KENT, P E. 1960. Geological results of petroleum exploration in Britain 1945–57. *Memoir of the Geological Society of London*, No. 2.

FEWTRELL, M D, and SMITH, D G. 1980. Revision of the Dinantian Stratigraphy of the Craven Basin, N England. *Geological Magazine*, Vol. 117, 37–49.

FLETCHER, T P. 1987. Geology of the Bashall Eaves area. *British Geological Survey Technical Report*, WA/87/44.

— 1990. Geology of the Whitewell area. *British Geological Survey Technical Report*, WA/90/53.

— 1991. Geology of the Dunsop Bridge area. *British Geological Survey Technical Report*, WA/91/80.

FORSTER, S C, and WARRINGTON, G. 1985. Geochronology of the Carboniferous, Permian and Triassic. 99–113 *in* The chronology of the geological record. SNELLING, N J (editor). *Memoir of the Geological Society of London*, No. 10.

FRASER, A J, NASH, D F, STEELE, R P, and EBDON, C C. 1990. A regional assessment of the intra-Carboniferous play of northern England. 417–439 *in* Classic petroleum provinces. BROOKS, J (editor). *Special Publication of the Geological Society of London*, No. 50.

GALE, S J. 1985. The Late- and Post-glacial environmental history of the southern Cumbrian massif and its surrounding lowlands. 282–297 *in The geomorphology of north-west England*. JOHNSON, R H (editor). (Manchester: Manchester University Press.)

GANDL, J. 1973. Die Karbon - Trilobiten des Kantabrischen Gebirges (NW-Spanien). I. Die Trilobiten der Vegamián-Schichten (Ober-Tournai). *Senckenbergiana Lethaia*, 54, S21-63, Taf. 4.

GASCOYNE, M, CURRANT, A P, and LORD, T C. 1981. Ipswichian fauna of Victoria Cave and the marine palaeoclimatic record. *Nature, London*, Vol. 294, 652–654.

GAWTHORPE, R L, and CLEMMEY, H. 1985. Geometry of submarine slides in the Bowland basin (Dinantian) and their relation to debris flows. *Journal of the Geological Society of London*, Vol. 142, 555–565.

— 1986. Sedimentation during carbonate ramp-to-slope evolution in a tectonically active area: Bowland Basin (Dinantian), northern England. *Sedimentology*, Vol. 33, 185–206.

— 1987. Tectono-sedimentary evolution of the Bowland Basin, northern England, during the Dinantian. *Journal of the Geological Society of London*, Vol 144, 58–71.

— GUTTERIDGE, P, and LEEDER, M R. 1989. Late Devonian and Dinantian basin evolution in northern England and North Wales. 1–23 *in* The role of tectonics in the Devonian and Carboniferous sedimentation in the British Isles. ARTHURTON R S, GUTTERIDGE, P, and NOLAN, S C (editors). *Yorkshire Geological Society Occasional Publication*, No. 6.

GEORGE, T N. 1978. Eustacy and tectonics: sedimentary rhythms and stratigraphical units in British Dinantian correlation. *Proceedings of the Yorkshire Geological Society*, Vol 42, 229–262.

— JOHNSON, G A L, MITCHELL, M, PRENTICE, J E, RAMSBOTTOM, W H C, SEVASTOPULO, G D, and WILSON, R B. 1976. A correlation of Dinantian rocks in the British Isles. *Special Report of the Geological Society of London*, No. 7.

GILL, M C. 1987. The Yorkshire and Lancashire lead mines: a study of lead mining in the south Craven and Rossendale districts. Northern Mine Research Society. *British Mining*, No. 33.

GRAYSON, R F, and OLDHAM, L. 1987. A new structural framework for the northern British Dinantian as a basis for oil, gas and mineral exploration. 33–59 *in European Dinantian environments*. MILLER, J, ADAMS, A E, and WRIGHT, V P (editors). (Chichester: John Wiley and Sons.)

GRESSWELL, R K. 1967. The geomorphology of the Fylde. 25–42 *in Liverpool essays in geography*. STEEL, R W, and LAWTON, R (editors). (London: Longmans.)

HALL, B R, and FOLLAND, C J. 1970. Soils of Lancashire. *Memoir of the Soil Survey of Great Britain*.

HAHN, G. 1966. Morphologie, Variabilität und postlarvale Ontogenie von *Archegonus (Phillibole) nitidus* und *Archegonus (Archegonus) winterbergensis* (Trilobita; unter-karbon). *Senckenbergiana Lethaia*, Vol 47, 347–383.

— and HAHN, R. 1971. Trilobiten aus dem unteren Teil det *crenistria*-Zone (unter-karbon, Cu III 1-2) des Rheinischen Schiefer-Gebirges. *Senckenbergiana Lethaia*, Vol. 52, 457–499.

— BRAUCKMANN, C, and SKALA, W. 1972. Kulm-Trilobiten aus der *striatus*-Zone Dinantium Cu IIIβ des Rheinischen Schiefer-Gebirges und des Harzes. *Senckenbergiana Lethaia*, Vol. 53, 31–63.

HARDING, T P. 1985. Seismic characteristics and identification of negative flower structures, and positive structural inversion. *Bulletin of the American Association of Petroleum Geologists*, Vol. 69, No. 4, 582–600.

HARRISON, D J. 1982. The limestone resources of the Craven Lowlands: description of parts of 1:50 000 Geological Sheets 59, 60, 61, 67, 68 and 69. *Mineral Assessment Report Institute of Geological Sciences*, No. 116.

HARVEY, A M, and RENWICK, W H. 1987. Holocene alluvial fan and terrace formation in the Bowland Fells, northwest England. *Earth Surface Processes and Landforms*, Vol. 12, 249–257.

HEALTH AND SAFETY EXECUTIVE. 1985. *The Abbeystead explosion.* (London: Her Majesty's Stationery Office.)

HIND, J W. 1907. On the occurrence of dendroid graptolites in British Carboniferous rocks. *Proceedings of the Yorkshire Geological Society*, Vol. 17, 97–109.

HIND, W, and HOWE, J A. 1901. The geological succession and palaeontology of the beds between the Millstone Grit and the Limestone-Massif at Pendle Hill and their equivalents in certain other parts of Britain. *Quarterly Journal of the Geological Society of London*, Vol. 57, 347–404.

HOLDSWORTH, B K, and COLLINSON, J D. 1988. Millstone Grit cyclicity revisited. 132–152 in *Sedimentation in a synorogenic basin complex; the Upper Carboniferous of Northwest Europe*. BESLEY, B M, and KELLING, G (editors). (Glasgow and London: Blackie.)

HOLLAND, C H (chairman), and others. 1978. A guide to stratigraphical procedure. *Special Report of the Geological Society of London*, No 11.

HOWARD, A S. In preparation Geology of the Bowland-with-Leagram and Bleasdale areas. *British Geological Survey Technical Report*, WA/87/45.

HUDSON, R G S. 1933. The scenery and geology of north-west Yorkshire. 228–255 *in* The geology of the Yorkshire Dales. *Proceedings of then Geologists' Association*, Vol. 44, 227–269.

— 1938. The general geology and the Carboniferous Rocks. 295–330 *in* The geology of the country around Harrogate. *Proceedings of the Geologists' Association*, Vol. 49, 295–352.

— 1944. A pre-Namurian fault-scarp at Malham. *Proceedings of the Leeds Philosophical Society* (Science section), Vol. 4, 226–232.

— and MITCHELL, G H. 1937. The Carboniferous geology of the Skipton Anticline. *Summary of Progress of the Geological Survey for 1935*, 1–45.

HUGHES, R A. 1986. Geology of the Trough of Bowland area (SD65SW). *British Geological Survey Technical Report*, WA/87/46.

— 1987. Geology of the Abbeystead area (SD 55 SE). *British Geological Survey Technical Report*, WA/87/41.

HULL, E. 1864. Geology of the country around Oldham, including Manchester and its suburbs. *Memoir of the Geological Survey of Great Britain.*

— DAKYNS, J R, TIDDEMAN, R H, WARD, J C, GUNN, W, and DE RANCE, C E. 1875. The geology of the Burnley Coalfield and of the country around Clitheroe, Blackburn, Preston, Chorley, Haslingden and Todmorden. *Memoir of the Geological Survey of Great Britain.*

JACKSON, D I, MULHOLLAND, P, JONES, S M, and WARRINGTON, G. 1987. The geological framework of the East Irish Sea Basin. 191–203 in *Petroleum geology of Northwest Europe*. BROOKS, J, and GLENNIE, K W (editors). (London: Graham and Trotman.)

JACKSON, J, and MCKENZIE, D. 1983. The geometrical evolution of normal fault systems. *Journal of Structural Geology*, Vol. 5, No. 5, 471–482.

JOHNSON, G A L. 1981. Geographical evolution from Laurasia to Pangaea. *Proceedings of the Yorkshire Geological Society*, Vol. 43, 221–252.

— and TARLING, D H. 1985. Continental convergence and closing seas during the Carboniferous. *Compte Rendu 10e Congrès International de Stratigraphie et de Géologie du Carbonifère (Madrid 1983)*, Vol. 4, 163–168.

JOHNSON, R H. 1985. The imprint of glaciation on the west Pennine uplands. 237–262 in *The geomorphology of north-west England*. JOHNSON, R H (editor). (Manchester: Manchester University Press.)

LAMB, H H. 1982. *Climate, history and the modern World.* (London: Methuen.)

LAWRENCE, S R, COSTER, P W, and IRELAND, R J. 1987. Structural development and petroleum potential of the northern flanks of the Bowland Basin (Carboniferous) north-west England. 225–233 in *Petroleum geology of north-west Europe*. BROOKS, J, and GLENNIE, K (editors). (London: Graham and Trotman.)

LEE, A G. 1988a. Studies of Carboniferous Basin configuration and evolution in northern and central England using gravity and magnetic data. Unpublished PhD thesis, University of Leeds.

— 1988b. Carboniferous basin configuration of central and northern England modelled using gravity data. 69–84 in *Sedimentation in a synorogenic basin complex; the Upper Carboniferous of Northwest Europe*. BESLEY, B M, and KELLING, G (editors). (Glasgow and London: Blackie.)

LEEDER, M R. 1982. Upper Palaeozoic basins of the British Isles - Caledonide inheritance versus Hercynian plate margin processes. *Journal of the Geological Society of London*, Vol. 139, 479–491.

— 1988. Recent developments in Carboniferous geology: a critical review with implications for the British Isles and NW Europe. *Proceedings of the Geologists' Association*, Vol. 99, 73–100.

LEES, A. 1964. The structure and origin of the Waulsortian (Lower Carboniferous) 'reefs' of west-central Eire. *Philosophical transactions of the Royal Society*, B247, 483–531.

— 1988. Waulsortian 'reefs': the history of a concept. *Mémoirs de l'Institut Gologique de l'Université de Louvain*, Vol. 34, 43–55.

— and MILLER, J. 1985. Facies variation in Waulsortian buildups, Part 2; Mid-Dinantian buildups from Europe and North America. *Geological Journal*, Vol. 20, No. 2, 159–180.

— HALLET, V, and HIBO, D. 1985. Facies variation in Waulsortian buildups, Part 1; A model from Belgium. *Geological Journal*, Vol. 20, No. 2, 133–158

LEVISON, A. 1988. The geology of the Morecambe gas field. *Geology Today*, Vol. 4, No. 3, 95–100.

LONGWORTH, D L. 1985. The Quaternary history of the Lancashire plain. 178–200 in *The geomorphology of north-west England*. JOHNSON, R H (editor). (Manchester: Manchester University Press.)

LOWE, D R. 1982. Sediment gravity flows: II. Depositional models with special reference to the deposits of high density turbidity currents. *Journal of Sedimentary Petrology*, Vol. 52, 279–297.

METCALFE, I. 1981. Conodont zonation and correlation of the Dinantian and early Numurian strata of the Craven lowlands of Northern England. *Report of the Institute of Geological Sciences*, No. 80/10, 1–70.

— RILEY, N J, and BURNETT, R D. (In press). The Craven Basin and adjacent areas. In *Conodont CAI patterns in the Carboniferous of Britain and Ireland*. BURNETT, R D, and HIGGINS, A C (editors).

MILLER, J. 1986. Facies relationships and diagenesis in Waulsortian mudmounds from the Lower Carboniferous of Ireland and N England. 311–335 in *Reef diagenesis*. SCHRODER, J H, and PURSER, B H (editors). (Berlin: Springer-Verlag.)

— and GRAYSON, R F, 1972. Origin and structure of the Lower Viséan "reef" limestones near Clitheroe, Lancashire. *Proceedings of the Yorkshire Geological Society*, Vol. 38, 607–638.

— — 1982. The regional context of Waulsortian facies in northern England. 17–33 in *Symposium on the paleoenvironmental setting and distribution of the Waulsortian facies*. BOLTON, K, LANE, R H, and LE MONE, D V (editors). (El Paso: El Paso Geological Society and the University of Texas.)

MOORE, E W J. 1930a. Species of the genus *Dimorphoceras* in the Bowland Shales. *Geological Magazine*, Vol. 67, 162–168.

— 1930b. A section in the Sabden Shales on the River Darwen, near Blackburn. *Journal of the Manchester Geologists Association*, Vol. 1, 103–108.

— 1936. The Bowland Shales from Pendle to Dinckley. *Journal of the Manchester Geologists Association*, Vol. 1, 167–192.

— 1939. The goniatite genus *Dimorphoceras* and its development in the British Carboniferous. *Proceedings of the Yorkshire Geological Society*, Vol. 24, 103–128.

— 1946. The Carboniferous goniatite genera *Girtyoceras* and *Eumorphoceras*. *Proceedings of the Yorkshire Geological Society*, Vol. 25, 387–445.

— 1950. The genus *Sudeticeras* and its distribution in Lancashire and Yorkshire. *Journal of the Manchester Geologists Association*, Vol. 2, 31–50.

— 1952. Notes on the genera *Prolecanites* and *Epicanites* with descriptions of two new species. *Liverpool and Manchester Geological Association*, Vol. 1, 71–76.

MOSELEY, F. 1954. The Namurian of the Lancaster Fells. *Quarterly Journal of the Geological Society of London*, Vol. 109, 423–54.

— 1956. The geology of the Keasden area, west of Settle, Yorkshire. *Proceedings of the Yorkshire Geological Society*, Vol. 30, 331–352.

— 1962. The structure of the south-western part of the Sykes Anticline, Bowland, West Yorkshire. *Proceedings of the Yorkshire Geological Society*, Vol. 33, 287–314.

NICOLAUS, H J. 1963. Zur stratigraphie und Fauna der *Crenistria* - Zone im Kulm des Rheinischer Schiefergebirges. *Beihefte zum Geologischen Jahrbuch*, Vol. 53, 1–246.

OAKES, D B, and SKINNER, A C. 1975. The Lancashire Conjunctive Use Scheme groundwater model. *Water Research Centre Technical Report*, TR 12.

OLDFIELD, F, and STATHAM, D C. 1965. Stratigraphy and pollen analysis on Cockerham and Pilling mosses, North Lancashire. *Memoirs and Proceedings of the Manchester Literary and Philosophical Society*, Vol. 107, 70–85.

PAPROTH, E, and 34 others. 1983. Bio-and lithostratigraphic subdivisions of the Dinantian in Belgium, a review. *Annales de la Societé Géologique de Belgique*, Vol. 106, 185–239.

PARKINSON, D. 1926. The faunal succession in the Carboniferous Limestone and Bowland Shales at Clitheroe and Pendle Hill, Lancashire. *Quarterly Journal of the Geological Society of London*, Vol. 82, 188–249.

— 1935. The geology and topography of the limestone knolls in Bolland (Bowland), Lancs. and Yorks. *Proceedings of the Geologists' Association*, Vol. 46, 97–120.

— 1936. The Carboniferous succession in the Slaidburn district, Yorkshire. *Quarterly Journal of the Geological Society of London*, Vol. 92, 294–331.

PEART, R J, BUSBY, J P, GREEN, C A, OGILVY, R D, and NANCARROW, P H A. 1989. Direct indication of hydrocarbons by airborne and ground magnetic survey. *British Geological Survey Technical Report*, WK/89/13R.

PHILLIPS, J. 1836. *Illustration of the geology of Yorkshire*. Part 2 *The Mountain Limestone District*. (London: John Murray.)

PLANT, J A, and JONES, D G (editors). 1989. *Metallogenic models and exploration criteria for buried carbonate-hosted ore deposits—a multidisciplinary study in eastern England*. (Keyworth, Nottingham: British Geological Survey. London: The Institution of Mining and Metallurgy.)

POWELL, C McA, and VEEVERS, J J. 1987. Namurian uplift in Australia and South America triggered the main Gondwanan glaciation. *Nature, London*, Vol. 326, No. 6109, 177–179.

PRICE, D, WRIGHT, W B, JONES, R C B, TONKS, L H, and WHITEHEAD, T H. 1963. Geology of the country around Preston. *Memoir of the Geological Survey of Great Britain*, Sheet 75 (England and Wales).

PRIOR, D B, BORNHOLD, B D, WISEMAN, W J, and LOWE, D R. 1987. Turbidity current activity in a British Columbia fjord. *Science*, Vol. 237, 1330–1333.

RAISTRICK, A. 1973. *Lead mining in the mid-Pennines*. (Truro: Bradford Barton.)

RAMSBOTTOM, W H C. 1973. Transgressions and regressions in the Dinantian: a new synthesis of British Dinantian stratigraphy. *Proceedings of the Yorkshire Geological Society*, Vol. 39, 567–607.

— 1974. Dinantian. 47–73 in *The geology and mineral resources of Yorkshire*. RAYNER, D H, and HEMINGWAY, J E (editors). (Yorkshire Geological Society.)

— RHYS, G H, and SMITH, E G. 1962. Boreholes in the Carboniferous rocks of the Ashover district, Derbyshire. *Bulletin of the Geological Survey of Great Britain*, No. 19, 75–168.

— and SAUNDERS, W B. 1985. Evolution and evolutionary biostratigraphy of Carboniferous ammonoids. *Journal of Paleontology*, Vol. 59, 123–139.

— CALVER, M A, EAGAR, R M C, HODSON, F, HOLLIDAY, D W, STUBBLEFIELD, C J, and WILSON, R B. 1978. A correlation of Silesian rocks in the British Isles. *Special Report of the Geological Society of London*, No. 10.

RAYNER, D H, and HEMINGWAY, J E (editors). 1974. *The geology and mineral resources of Yorkshire*. (Leeds: Yorkshire Geological Society.)

RILEY, N J. 1981. Field meetings 1980. The Carboniferous of Slaidburn, Yorkshire, 7th June 1980. *Proceedings of the Yorkshire Geological Society*, Vol. 43, 467.

— 1982. Early Viséan trilobite and ammonoid faunas in the western part of the Craven Basin. Unpublished PhD. thesis, University of Bristol.

— 1985. *Asturoceras* and other dimorphoceratid ammonoids from the Namurian (E_2) of Lancashire. *Proceedings of the Yorkshire Geological Society*, Vol. 45, 219–224.

— 1987. Type ammonoids from the Mid-Carboniferous boundary interval in Britain. *Courier Forschungsinstitut Senckenberg*, Vol. 98, 25–37.

— 1990a. Revision of the *Beyrichoceras* Ammonoid-Biozone (Dinantian), NW Europe. *Newsletters in Stratigraphy*, Vol. 21, 149–156.

— 1990b. Stratigraphy of the Worston Shale Group (Dinantian), Craven Basin, NW England. *Proceedings of the Yorkshire Geological Society*, Vol. 48, 163–189.

— 1991. A global review of mid-Dinantian ammonoid zonation. *Courier Forschungsinstitut Senckenberg*.

ROSE, W C C, and DUNHAM, K C. 1977. Geology and hematite deposits of South Cumbria. *Economic Memoir of the Geological Survey of Great Britain*, Sheets 58, part 48.

ROWLEY, D B, RAYMOND, A, PARRISH, J T, LOTTES, A, SCOTESE, C R, and ZEIGLER, A M. 1985. Carboniferous paleogeographic, phytogeographic and paleoclimatic reconstructions. *International Journal of Coal Geology*, Vol. 5, 7–42.

SAGE, R C, and LLOYD, J W. 1978. Drift deposit influences on the Triassic Sandstone aquifer of NW Lancashire as inferred by hydrochemistry. *Quarterly Journal of Engineering Geology*, Vol. 11, 209–218.

SEVASTOPULO, G D. 1981. Lower Carboniferous. 147–171 in *A geology of Ireland*. HOLLAND, C H (editor). (Edinburgh: Scottish Academic Press.)

SIMS, A P. 1988. The evolution of a sand-rich basin-fill sequence in the Pendleian (Namurian, E1c) of north-west England. Unpublished PhD thesis, University of Leeds.

SIMPSON, J, and KALVODA, J. 1987. Sedimentology and foraminiferal biostratigraphy of the Arundian (Dinantian) stratotype. 226–237 in *Micropalaeontology of carbonate environments*. HART, M B (editor). *British Micropalaeontological Society Series*. (Chichester: Ellis Horwood.)

SMITH, D B, BRUNSTROM, R G W, MANNING, P I, SIMPSON, S, and SHOTTON, F W. 1974. A correlation of Permian rocks in the British Isles. *Special Report of the Geological Society of London*, No. 5.

SMITH, K, SMITH, N J P, and HOLLIDAY, D W. 1985. The deep structure of Derbyshire. *Geological Journal*, Vol. 20, 215–225.

STEPHENS, J V, MITCHELL, G H, and EDWARDS, W. 1953. Geology of the country between Bradford and Skipton. *Memoir of the Geological Survey of Great Britain*, Sheet 69 (England and Wales).

STOW, D, and SHANMUGAM, G. 1980. Sequence of structures in fine-grained turbidites: comparison of recent deep-sea and ancient flysh sediments. *Sedimentary Geology*, Vol. 5, 23–42.

TALBOT, M S. 1974. *The Chipping gap: stone counts and till fabric analysis*. Unpublished BEd dissertation, Chorley College of Education.

TALLIS, J H. 1985. Erosion of blanket peat in the southern Pennines: new light on an old problem. 313–336 in *The geomorphology of north-west England*. JOHNSON, R H (editor). (Manchester: Manchester University Press).

TAYLOR, T L. 1961. *Geomorphology of Over Wyre*. Unpublished MA thesis, University of Liverpool.

THOMPSON, D B. 1970. The stratigraphy of the so-called Keuper Sandstone Formation (Scythian–?Anisian) in the Permo-Triassic Cheshire Basin. *Quarterly Journal of the Geological Society of London*, Vol. 126, 151–181.

TIDDEMAN, R H. 1872. On the evidence for the ice-sheet in North Lancashire and adjacent parts of Yorkshire and Westmoreland. *Quarterly Journal of the Geological Society of London*, Vol. 28, 471–491.

— 1889. On concurrent faulting and deposit in Carboniferous times in Craven, Yorkshire, with a note on Carboniferous reefs. *Report of the British Association* (Newcastle), 600–603.

TISSOT, B P, and WELTE, D H. 1984. *Petroleum formation and occurrence* (2nd edition, revised). (Berlin, Heidelberg, New York, Tokyo: Springer-Verlag.)

TOOLEY, M J. 1980. Theories of coastal change in north-west England. 74–86 in *Archaeology and coastal change*. THOMPSON, F H (editor). (The Society of Antiquaries of London.)

TREWIN, N H. 1968. Potassium bentonites in the Namurian of Staffordshire and Derbyshire. *Proceedings of the Yorkshire Geological Society*, Vol. 37, 73–91.

VARKER, W J, and SEVASTOPULO, G D S. 1985. The Carboniferous System: Part 1—Conodonts of the Dinantian Subsystem from Great Britain and Ireland. 167–209 in *A stratigraphical index of conodonts*. HIGGINS, A C, and AUSTIN, R L (editors). *British Micropalaeontological Society Series*. (Chichester: Ellis Horwood).

WADGE, A J, BATESON, J H, and EVANS, A D. 1983. Mineral reconnaissance surveys in the Craven Basin. *Mineral Reconnaissance Programme Report, Institute of Geological Sciences*, No. 66.

WALKER, C T. 1967. Comparison of upper Viséan sedimentology to the Bowland Shale overlap in Yorkshire, England. *Sedimentary Geology*, Vol. 1, 117–136.

WARRINGTON, G, AUDLEY-CHARLES, M G, ELLIOTT, R E, EVANS, W B, IVIMEY-COOK, H C, KENT, P E, ROBINSON, P L, SHOTTON, F W, and TAYLOR, F M. 1980. A correlation of the Triassic rocks in the British Isles. *Special Report of the Geological Society of London*, No. 13.

WELD, A. 1851. An account of a remarkable flood at Chipping in Lancashire. *The London, Edinburgh and Dublin Philosophical Magazine and Journal of Science*, Vol. 2, No. 8, 209–215.

WILSON, A A. 1990. The Mercia Mudstones Group (Trias) of the East Irish Sea Basin. *Proceedings of the Yorkshire Geological Society*, Vol. 48, 1–22.

— BRANDON, A, and JOHNSON, E W. 1989. Geological context of the Wyresdale Tunnel methane explosion. *British Geological Survey Technical Report*, WA/89/30.

— and EVANS, W B. 1990. Geology of the country around Blackpool. *Memoir of the British Geological Survey*, Sheet 66 (England and Wales).

WINTERS, W A. 1961. Landforms associated with stagnant ice. *Professional Geographer*, Vol. 13, 19–23.

WORTHINGTON, P F. 1972. A geoelectrical investigation of the drift deposits in north-west Lancashire. *Geological Journal*, Vol. 8, 1–16.

— 1977. Permeation properties of the Bunter Sandstone of northwest Lancashire, England. *Journal of Hydrology*, Vol. 32, 295-303.

— and GRIFFITHS, D H. 1975. The application of geophysical methods in the exploration and development of sandstone aquifers. *Quarterly Journal of Engineering Geology*, Vol. 8, 73–102.

APPENDIX 1

Selected boreholes

This list includes the permanent record number, location, total depth and stratigraphical range of the selected boreholes that are referred to in this memoir. The number given by the original owner of the borehole is also shown in parenthesis where appropriate. The sites and brief abstract logs of most of the holes listed are shown on the relevant 1:10 000 geological maps listed in Appendix 1. Copies of these records may be obtained from the British Geological Survey, Keyworth, Nottingham NG12 5GG at a fixed tariff. Other nonconfidential borehole records are held on open file in the Survey's archives.

SD 33 NE/17 Thistleton [3976 3700] details 'commercial in confidence'.

SD 43 NW/1 Thistleton Bridge [4079 3836] 129.9 m Singleton and Kirkham mudstone formations, and drift.

SD 43 NW/2 White Carr Farm [4343 3527] 6.7 m drift.

SD 43 NE/1a Inskip Creamery [4628 3769] 24.4 m Singleton Mudstone Formation, drift.

SD 43 NE/1b Inskip Creamery [4628 3769] 161.5 m Sherwood Sandstone Group, Hambleton Mudstone Formation, Singleton Mudstone Formation and drift.

SD 43 NE/2 Woodsfold Bridge [4759 3683] 106.8 m Sherwood Sandstone Group, ?Hambleton Mudstone Formation and drift.

SD 43 NE/4 Sowerby Hall Farm [4749 3869] 92.3 m Singleton Mudstone Formation and drift.

SD 43 SW/6 Kirkham [4324 3247] 444.93 m Sherwood Sandstone Group, Hambleton Mudstone Formation, Singleton Mudstone Formation, Kirkham Mudstone Formation, Breckells Mudstone Formation and drift.

SD 43 SE/6 (Fylde Water Board FF1) [4845 3463] Sherwood Sandstone Group and drift.

SD 44 NW/2 (ICI No.16A) Tarn Farm [4377 4532] 131.0 m Sherwood Sandstone Group and drift.

SD 44 NW/4 (ICI No.15) North Wood's Hill Farm [4475 4552] 161.85 m Sherwood Sandstone Group and drift.

SD 44 NW/7 (ICI No.11) Bone Hill Farm [4371 4637] 132.7 m Sherwood Sandstone Group and drift.

SD 44 NW/9 (Fylde Water Board T18) Moss Edge Farm [4397 4926] 44.8 m Sherwood Sandstone Group and drift.

SD 44 NW/12 (ICI Observation Well A) Pilling [4042 4833] 30.0 m Sherwood Sandstone Group and drift.

SD 44 NE/2 (Fylde Water Board T8) Broad Lane [4818 4687] 91.4 m Sherwood Sandstone Group and drift.

SD 44 NE/7 (Fylde Water Board T12) Gubberford Lane [4992 4795] 37.5 m Sherwood Sandstone Group and drift.

SD 44 NE/9 (Fylde Water Board T10) [4931 4590] 42.8 m Sherwood Sandstone Group and drift.

SD 44 NE/13 (Fylde Water Board T46) [4909 4990] Westfield 40.2 m Manchester Marls, Sherwood Sandstone Group and drift.

SD 44 SW/1 Wall Pool Bridge [4116 4044] 121.9 m Singleton Mudstone Formation and drift.

SD 44 SW/2 (ICI Observation Well N) [4220 4329] Crook Farm 56.4 m Sherwood Sandstone Group, Hambleton Mudstone Formation, Singleton Mudstone Formation and drift.

SD 44 SW/5 South Wood's Hill Farm [4385 4475] 8.86 m drift.

SD 44 SE/5 (Fylde Water Board T15) Sharples Lane [4299 4695] 50.0 m Sherwood Sandstone Group and drift.

SD 44 SE/11 (ICI Observation Well P) Moss Side [4581 4220] 91.4 m Singleton Mudstone Formation and drift.

SD 44 SE/24 (Fylde Water Board T51) Primrose Hill [4600 4499] 46.6 m drift.

SD 44 SE/27 (Lancashire River Authority T55) [4971 4158] 31.7 m Sherwood Sandstone Group and drift.

SD 44 SE/28 (Lancashire River Authority T69) Myerscough [4985 4105] 30.0 m Sherwood Sandstone Group and drift.

SD 44 SE/29 (Fylde Water Board M4) Hamer Field Wood [4940 4234] 153.4 m Sherwood Sandstone Group and drift.

SD 44 SE/58 (Longtown to Warburton Gas Pipeline No.116) Manor House Farm [4779 4010] 8.0 m drift.

SD 53 NW/1 (Benson's) [5398 3643] 90.83 m Worston Shale Group and drift.

SD 53 NW/4 (Fylde Water Board F1) Barton Mill [5246 3704] 75.9 m Lower Bowland Shale Formation, Manchester Marls, Sherwood Sandstone Group and drift.

SD 53 NW/5 (Fylde Water Board H1) White Horse Lane [504 382] ?Manchester Marls, Sherwood Sandstone Group and drift.

SD 53 SW/7 (Fylde Water Board C1) [5192 3486] 193.7 m ?Pendleside Limestone Formation, ?Manchester Marls, Sherwood Sandstone Group and drift.

SD 53 SE/12 (Courtaulds A) [5756 3292] 85.34 m Sherwood Sandstone Group and drift.

SD 53 SE/33 (M6 Bridgeworks No.5258/1) [5582 3326] 7.6 m drift.

SD 53 SE/35 (M6 Savick Brook No.2) [5575 3328] 7.6 m drift.

SD 54 NW/1 Oakenclough [5371 4769] 213.36 m Pendle Grit Formation and drift.

SD 54 NW/2 Oakenclough [5478 4726] 122.22 m Pendle Grit Formation and drift.

SD 54 NW/3 (Fylde Water Board T44) Parkhead Bridge [5061 4552] 31.09 m Manchester Marls and drift.

SD 54 NW/4 (Fylde Water Board T44A) Disused railway cutting [5046 4540] 22.25 m Sherwood Sandstone Group.

SD 54 SW/1 Garstang Creamery [5095 4409] 34.75 m ?Roeburndale Formation, Manchester Marls and drift.

SD 54 SW/2 (Fylde Water Board K1) [5005 4009] 152.4 m Manchester Marls, Sherwood Sandstone Group and drift.

SD 54 SW/5A (Fylde Water Board R3) [5110 4282] 148.74 m Manchester Marls, Sherwood Sandstone Group and drift.

SD 54 SW/6 (Fylde Water Board R1) Stubbins Bridge [5090 4290] 139.60 m Manchester Marls, Sherwood Sandstone Group and drift.

SD 54 SW/7 (Fylde Water Board N1) Brock [5120 4055] 108.20 m Manchester Marls, Sherwood Sandstone Group and drift.

SD 54 SW/9 (Fylde Water Board P1) [5030 4209] 146.30 m Manchester Marls, Sherwood Sandstone Group and drift.

SD 54 SW/7   (Fylde Water Board N1) Brock [5120 4055] 108.20 m Manchester Marls, Sherwood Sandstone Group and drift.

SD 54 SW/9   (Fylde Water Board P1) [5030 4209] 146.30 m Manchester Marls, Sherwood Sandstone Group and drift.

SD 54 SW/13   (Fylde Water Board T40) [5059 4034] 39.62 m Sherwood Sandstone Group and drift.

SD 54 SW/17   (Lancashire River Authority T59) [5266 4346] 46.33 m Upper Bowland Shale Formation including C.malhamense Marine Band, and drift.

SD 54 SW/19   (Lancashire River Authority T61) [5182 4332] 36.58 m ?Roeburndale Formation and drift.

SD 54 SW/21   (Lancashire River Authority T58) [5127 4254] 51.82 m Manchester Marls and drift.

SD 54 SW/23   (Lancashire River Authority T57) Ducketts Farm [5191 4137] 34.74 m Pendle Grit Formation and drift.

SD 54 SW/29   (M6 No.5267/1) Footbridge at River Calder [5116 4345] 6.4 m drift.

SD 55 SW/9   (M6 No.5286/2) Nan's Nook Footbridge [5073 5142] 5.79 m Wellington Crag Sandstone and drift.

SD 64 NW/2   (BP Minerals International MHD 7) Dinkling Green Farm [6434 4718] 200.55 m Clitheroe Limestone Formation.

SD 64 NE/1   (BGS Cow Ark No.1) [6802 4677] 32.05 m Clitheroe Limestone Formation, Hodder Mudstone Formation and drift.

SD 64 NE/2   (BGS Cow Ark No.2) [6802 4677] 67.90 m (inclined) Clitheroe Limestone Formation, Hodder Mudstone Formation and drift.

SD 64 NE/3   (BGS Cow Ark No.3) [6798 4677] 33.89 m (inclined) Clitheroe Limestone Formation, Hodder Mudstone Formation and drift.

SD 64 NE/4   (BGS Cow Ark No.4) [6790 4678] 100.92 m Clitheroe Limestone Formation, Hodder Mudstone Formation.

SD 64 NE/5   (BGS Cow Ark No.5) [6788 4669] 120.65 m Clitheroe Limestone Formation, Hodder Mudstone Formation, Hodderense Limestone Formation, Pendleside Limestone Formation.

SD 64 NE/6   (BGS Cow Ark No.6) [6780 4661] 29.00 m Pendleside Limestone Formation, Lower Bowland Shale Formation and drift.

SD 64 NE/7   (BGS Cow Ark No.7) [6781 4678] 222.0 m (inclined) Chatburn Limestone Group, Clitheroe Limestone Formation, Hodder Mudstone Formation, Hodderense Limestone Formation, Pendleside Limestone Formation, Lower Bowland Shale Formation and drift.

SD 64 NE/9   (BGS Cow Ark No.9) [6832 4647] 260 m (inclined) Chatburn Limestone Formation, Hodder Mudstone Formation, Hodderense Limestone Formation, Pendleside Limestone Formation, Lower Bowland Shale Formation and drift.

SD 64 NE/10   (BGS Cow Ark No.10) [6807 4665] c.257 m (inclined) Chatburn Limestone Formation, Clitheroe Limestone Formation, Hodder Mudstone Formation and drift.

SD 64 NE/14   (BP Minerals International MHD 1) [6816 4665] 399.43 m Clitheroe Limestone Formation, Hodder Mudstone Formation and drift.

SD 64 NE/16   (BP Minerals International MHD 3) [6831 4625] 274.04 m Clitheroe Limestone Formation, Hodder Mudstone Formation, Hodderense Limestone Formation, Pendleside Limestone Formation, Lower Bowland Shale Formation and drift.

SD 64 NE/17   (BP Minerals International MHD 4) [6742 4675] 149.51 m Clitheroe Limestone Formation, Hodder Mudstone Formation, Lower Bowland Shale Formation, Upper Bowland Shale Formation and drift.

SD 64 NE/18   (BP Mineral International MHD 5) [6666 4700] 150.98 m Hodder Mudstone Formation and drift.

SD 64 NE/19   (BP Minerals International MHD 8) [6771 4662] 167.41 m Hodder Mudstone Formation, Hodderense Limestone Formation, Pendleside Limestone Formation, Lower Bowland Shale Formation and drift.

SD 64 NE/20   (BP Minerals International MHD 11) [6833 4683] 300.58 m Hodder Mudstone Formation, Hodderense Limestone Formation, Pendleside Limestone Formation, Lower Bowland Shale Formation, Upper Bowland Shale Formation and drift.

SD 64 NE/23   (BP Minerals International MHD 18) [6823 4643] 270.39 m Hodder Mudstone Formation, Hodderense Limestone Formation, Pendleside Limestone Formation and Lower Bowland Shale Formation.

APPENDIX 2

Petrographical data

List of petrographical samples

E numbers refer to the Sliced Rock Collection of the British Geological Survey. Each number is followed by locality details; in the case of records in the Survey's borehole archives, the localities are indicated by borehole reference numbers (*see* Appendix 1).

E41084 [SD 54 SW/21]; borehole (L.R.A. T58), depth 40.23 m. Manchester Marl.

E41085 [SD 54 SW/21]; borehole (L.R.A. T58), depth 48.08 m. Manchester Marl.

E61937 Ditch [6065 4114] near Crow Trees Farm; Buckbanks Sandstone.

E61938 Disused Quarry [5816 4372] near Lower Core (Plate 8); Pendleside Sandstones.

E61939 Stream [5938 4253] near Blacksticks Farm; Pendleside Sandstones.

E61940 Disused Quarry [6071 4359] Lingey Hill, Chipping; Pendleside Sandstones.

E61941 At waterfall in River Brock [5322 4134] below Walmsley Bridge; Upper Bowland Shales.

E61942 Gully [5985 4461] on south face of Parlick; Upper Bowland Shales.

E61943 Disused Quarry [6365 4013] near Cardwell House; Pendle Grit, basal part, coarse-grained facies.

E61944 Disused Quarry [6365 4013] near Cardwell House; Pendle Grit, basal part, fine-grained facies.

E61945 River Calder [5188 4350] immediately below Sandholme Bridge; Pendle Grit, middle part.

E61946 Grizedale Brook [5140 4800]; Pendle Grit c.8.0 m below the top of the formation.

E61947 Stream [5116 4603] at All Saints Church near Barnacre Lodge; sandstone immediately below E.ferrimontanum Marine Band.

E61948 Stream [5116 4603] at All Saints Church near Barnacre Lodge; sandstone immediately above E.ferrimontanum Marine Band (Plate 13).

E61949 Disused Quarry [5068 4940] in Park Wood, near Wyresdale Park; Park Wood Sandstones.

Analyses

The results of a petrographical analysis of samples E61937 to E61949 (inclusive) made by Mr B Humphreys are given in Tables 3, 4, and 5 below.

Table 3 Detrital mineralogy of sandstones.

Sample	%Qm	Qp	L	K	P	M	DC	CEM
E61937	80.8	2.6	0.6	—	—	0.6	2.8	12.6
E61938	80.2	1.8	1.8	—	2.8	2.0	7.0	4.4
E61939	83.8	1.6	2.2	—	1.6	0.6	5.8	4.4
E61940	83.2	1.4	1.8	—	3.2	2.6	3.8	4.0
E61943	59.2	8.8	1.0	12.8	4.2	2.0	8.6	3.4
E61944	71.2	2.4	2.4	2.6	4.8	4.6	4.8	7.2
E61945	78.4	2.0	0.6	3.6	4.8	2.4	3.0	5.2
E61946	75.2	1.8	0.4	4.8	6.0	2.8	2.4	6.6
E61947	60.8	1.2	1.4	6.6	3.0	1.0	2.4	23.6
E61948	56.0	0.6	1.4	5.6	4.8	4.0	3.6	24.0
E61949	71.2	5.8	1.6	5.0	2.6	2.4	3.0	8.4

Percentages stated are deduced from thin-section point counts of 500 grains.

QM = Monocrystalline quartz
Qp = Polycrystalline quartz
L = Lithics (sedimentary & metamorphic rock fragments)
K = K-feldspars
P = Plagioclase feldspars (albite)
M = Mica flakes (muscovite and biotite)
DC = Detrital clay (pore-filling matrix or thin clay laminae)
CEM = Total cement present. For details see Table 4.

Table 4   Summary of grain size and porosity data.

Sample	Grain size (modal size from disaggregated sandstones)	Porosity	Principal cements (% of total mineralogy)
E61937	Fine sand	7.6%	Quartz overgrowths – 4.8% Kaolinite – 3.6% Chlorite – 2.6% Iron oxides – 1.6%
E61938	Fine sand	2.7%	Pyrite – 1.4% Calcite – 1.0% Quartz overgrowths – 1.0% Illite – 0.6% Kaolinite – 0.4%
E61939	Very fine to fine sand	6.0%	Quartz overgrowths – 3.2% Kaolinite – 0.8% Iron oxides – 0.4%
E61940	Fine sand	4.9%	Quartz overgrowths – 1.8% Kaolinite – 1.0% Illite – 0.6% Pyrite – 0.6%
E61941	Carbonate-cemented ?spiculite	—	Calcite and dolomite >50%
E61942	Dolomite-cemented ?fine sand	—	Ferroan dolomite with minor calcite >40%
E61943	Coarse sand	3.8%	Kaolinite – 1.4% Illite – 0.8% Quartz overgrowths – 0.6% Iron oxides – 0.6%
E61944	Very fine to fine sand	4.9%	Kaolinite – 2.8% Quartz overgrowths – 2.4% Iron oxides – 1.6% ?Siderite – 0.4%
E61945	Fine sand	4.8%	Kaolinite – 2.8% Quartz overgrowths – 1.4% Illite – 1.0%
E61947	Fine to medium sand	0.6%	Non-ferroan calcite & dolomite – 16.2% Ferroan dolomite – 2.6% Quartz overgrowths – 2.2% ?Siderite – 1.4% Kaolinite – 0.8% Iron oxides – 0.4%
E61948	Fine sand	not measurable	Ferroan calcite – 19.4% Iron oxides – 2.4% ?Siderite – 1.2% Kaolinite – 0.6% Quartz overgrowths – 0.4%
E61949	Fine sand	7.7%	Quartz overgrowths – 3.8% Illite – 2.6% Kaolinite – 1.6% Iron oxides – 0.4%

Percentages stated are deduced from thin-section point counts of 500 grains.

Table 5   Clay mineralogy of the $<2\,\mu$ fraction.

E61937	Chlorite + Illite + Mixed-layer illite-smectite + (Kaolinite)
E61938	Illite (poorly crystalline) + ?ordered I/S + Kaolinite
E61939	Illite + ordered I/S + Kaolinite
E61940	Illite
E61941	No clay minerals detected
E61942	(Kaolinite)
E61943	Kaolinite + Chlorite + Illite
E61944	Kaolinite + Chlorite + Illite
E61945	Kaolinite + Illite + (randomly interstratified I/S)
E61946	Kaolinite + Illite + randomly interstratified I/S
E61947	Kaolinite + (Illite) + (Chlorite)
E61948	Kaolinite + (Chlorite)

The dominant clay phase is listed first, with subordinate phases listed in order of decreasing abundance.

Brackets indicate minor amounts.

The name "illite" is used as a nonspecific term to indicate a discrete phase of clay-grade mica which may include minor amounts of interstratified smectite (i.e. traces of ordered mixed layer illite-smectite). Where mixed-layer illite smectites (I/S) are a significant component of the assemblage, the term "randomly interstratified I/S" is used when the smectite component expands to c.17.7Å with glycerolation, and "ordered I/S" when full expansion of the smectite component is not observed-in such cases the smectite component is generally less than 40% of the mixed-layer clay.

Chlorite in sample E61937: the 14Å reflection is stronger than the 7Å reflection suggesting a Mg-rich variety of chlorite. The 14Å reflection does not expand with glycerolation, but collapses from c.14.2Å to c.13.2Å with heating to 600°C suggesting interstratification of small amounts of a collapsible phase such as smectite or vermiculite.

APPENDIX 3

Open-file reports

Open-file reports containing geological details additional to those shown on the 1:10 000 maps are listed below. They can be consulted at BGS libraries or purchased from the same outlets as the dyeline maps.

SD 44 NW	(Pilling)	RGC	WA/87/34
SD 44 NE	(Garstang)	RGC	WA/87/35
SD 44 SW	(Out Radcliffe and Great Ecclestone)	RGC	WA/87/36
SD 44 SE	(St Michael's on Wyre)	RGC	WA/87/37
SD 45 SW	(Cockersand Abbey)	RGC	WA/87/38
SD 45 SE	(Cockerham)	RGC	WA/87/39
SD 53 NW & SW (part)	(Barton and Fulwood)	DMcCB	WA/87/40
SD 53 NE & NW (part)	(Goosnargh and north-east Preston)	DMcCB	WA/87/42
SD 54 SE	(Beacon Fell)	NA	WA/89/33
SD 55 SE	(Abbeystead)	RAH	WA/87/41
SD 63 NW & SW (part)	(Longridge and Ribchester)	DMcCB	WA/89/66
SD 63 NE & SE (part)	(Hurst Green and Wilpshire)	DMcCB	WA/87/43
SD 64 NW	(Bowland with Leagram)	ASH	WA/87/45
SD 64 NE	(Whitewell)	TPF	WA/90/53
SD 64 SW	(Chipping)	NA	WA/90/35
SD 64 SE	(Bashall Eaves)	TPF	WA/87/44
SD 65 SW	(Trough of Bowland)	RAH	WA/87/46
SD 65 SE	(Dunsop Bridge)	TPF	WA/91/80

APPENDIX 4

Geological Survey photographs

Seventy two photographs illustrating the geology of the Garstang district are deposited for reference in the headquarters library of the British Geological Survey, Keyworth, Nottingham NG12 5GG; in the library at the BGS, Murchison House, West Mains Road, Edinburgh EH9 3LA; and in the BGS Information Office at the Natural History Museum Earth Galleries, Exhibition Road, London SW7 2DE. They belong to the A Series and were taken either in March 1986 or September 1988 during and following the survey. The photographs depict details of the various rocks and sediments exposed and also include general views and scenery. A list of titles can be supplied on request. The photographs can be supplied as black and white or colour prints and 2 × 2 colour transparencies, at a fixed tariff.

FOSSIL INDEX

Page numbers in italics refer to figures.
Page numbers in bold refer to tables.
Page numbers followed by 'P' refer to plates.

GENERAL INDEX

Page numbers in italics refer to figures.
Page numbers in bold refer to tables.
Page numbers followed by 'P' refer to plates.

BRITISH GEOLOGICAL SURVEY

Keyworth, Nottingham NG12 5GG
0602-363100

Murchison House, West Mains Road,
Edinburgh EH9 3LA   031-667 1000

London Information Office, Natural History Museum Earth
Galleries, Exhibition Road, London SW7 2DE
071-589 4090

The full range of Survey publications is available through the
Sales Desks at Keyworth and at Murchison House,
Edinburgh, and in the BGS London Information Office in the
Natural History Museum Earth Galleries. The adjacent
bookshop stocks the more popular books for sale over the
counter. Most BGS books and reports are listed in HMSO's
Sectional List 45, and can be bought from HMSO and
through HMSO agents and retailers. Maps are listed in the
BGS Map Catalogue, and can be bought from Ordnance
Survey agents as well as from BGS.

*The British Geological Survey carries out the geological survey of Great
Britain and Northern Ireland (the latter as an agency service for the
government of Northern Ireland), and of the surrounding continental
shelf, as well as its basic research projects. It also undertakes
programmes of British technical aid in geology in developing countries as
arranged by the Overseas Development Administration.*

*The British Geological Survey is a component body of the Natural
Environment Research Council.*

Maps and diagrams in this book use topography based on
Ordnance Survey mapping

HMSO publications are available from:

HMSO Publications Centre
(Mail, fax and telephone orders only)
PO Box 276, London, SW8 5DT
Telephone orders 071-873 9090
General enquiries 071-873 0011
(queuing system in operation for both numbers)
Fax orders 071-873 8200

HMSO Bookshops
49 High Holborn, London, WC1V 6HB
071-873 0011 Fax 071-873 8200 (counter service only)
258 Broad Street, Birmingham, B1 2HE
021-643 3740 Fax 021-643 6510
Southey House, 33 Wine Street, Bristol, BS1 2BQ
0272 264306 Fax 0272 294515
9-21 Princess Street, Manchester, M60 8AS
061-834 7201 Fax 061-833 0634
16 Arthur Street, Belfast, BT1 4GD
0232 238451 Fax 0232 235401
71 Lothian Road, Edinburgh, EH3 9AZ
031-228 4181 Fax 031-229 2734

HMSO's Accredited Agents
(see Yellow Pages)

and through good booksellers